BEEKEEPING WITH WOVEN HIVES IN GREECE THROUGH THE AGES

GEORGIOS MAVROFRIDIS

Beekeeping With Woven Hives In Greece Through The Ages
Copyright © 2023 Georgios Mavrofridis

All rights reserved. No part of this publication may be reproduced, stored in a retrieval system, transmitted in any form or by any means electronic, mechanical, including photocopying, recording or otherwise without prior consent of the copyright holders.

ISBN 978-1-914934-61-2

Published by Northern Bee Books, 2023
Scout Bottom Farm
Mytholmroyd
Hebden Bridge HX7 5JS (UK)

Design and artwork by DM Design and Print

CONTENTS

	Page
ACKNOWLEDGMENTS	v
INTRODUCTION	1
1. WICKER-HIVE TYPES AND THE PRACTICE OF BEEKEEPING	3
1.1 Upright Wicker Hives	3
1. 1. 1 Open-at-the-top and open-at-the-top-and-bottom movable-comb wicker hives	3
1. 1. 2 Skeps	37
1. 1. 2. 1 *The migratory skep of northern Greece*	37
1. 1. 2. 2 *The migratory skep of Attica and Boeotia*	58
1. 1. 2. 3 *Conical and bell-shaped skeps*	61
1.2 Horizontal Wicker Hives	75
2. CONSTRUCTION OF WOVEN HIVES	80
3. WOVEN HIVES IN ANTIQUITY AND THE MIDDLE AGES	93
4. THE INFLUENCE OF GREEK WOVEN HIVES ON THE EVOLUTION OF WORLD BEEKEEPING	100
EPILOGUE	108
NOTES	109
BIBLIOGRAPHY	128

The Eva Crane Trust

www.evacranetrust.org

The Trust was formed by Dr Eva Crane in 2002. Its aim is to advance the understanding of bees and bee science including the history and development of beekeeping. It does this by helping to fund the collection, collation, dissemination, and publication of research worldwide as well as supporting projects to record and propagate further understanding of beekeeping practices through historical and contemporary discoveries. The Trust is, therefore, delighted to have facilitated the publication of this excellent record by Georgios Mavrofridis which it sees as a direct and continuous extension of the work and aims of its Founder.

Richard Jones
Chairman of the Eva Crane Trust
June 2023

ACKNOWLEDGMENTS

The present study could not have been completed without the help and assistance of several people, who in one way or another contributed to its realization. I must, therefore, express my warmest thanks: to the Hellenic Folklore Research Centre of the Academy of Athens, and personally to its Director, Dr. Evangelos Karamanes, for permission to study manuscripts at the Centre's Archives; to the Centre for Research in Modern-Greek Dialects and Idioms of the Academy of Athens, for permission to access precious manuscripts at the Centre's Archives; to the Institute of Agricultural Sciences and personally to its Director, Dr. Georgios Balotis, for permission to study the exhibits of the Beekeeping Collection held there, as well as for the use of material from the Institute's Archives; to the Hellenic Scientific Society of Apiculture – Sericulture, and personally to its president, Dr. Sofia Gounari, for permission to dive into the archives of Thanasis Bikos and Penelope (Poppy) Papadopoulo in its possession; to the Basketry Museum of Komotini, and personally to its Director, Dr. Antonis Liapis, for the valuable information, photographs and general help; to the Folklore Museum of Didymoteicho, and personally to its Director Mrs. Chrysoula Tsakiraki-Kyroudi, for the information and permission to publish photographs of the Museum's exhibits; to the Ethnological Museum of Thrace, and personally to its Director, Mrs. Angeliki Giannakidou, for the information regarding some of the exhibits; to the Museum of Folklore and Culture of Florina, and especially to the agronomist Ioannis Anagnostopoulos, for the information and license to use photographs from the Museum's Beekeeping Collection; to the Eva Crane Trust for permission to use photographs from its archive, to the late agronomist and researcher of traditional Greek beekeeping, Thanasis Bikos (1948-2015), for his valuable information during our numerous discussions between 2006 and 2014; to the Professor of History of Modern Hellenism at the University of Athens, Dr. Maria Efthymiou for permission to use images and drawings from her articles in the *Ethnographica* journal; to the former Professor of Beekeeping and Honey Bee Pathology at the University of Thessaloniki, Dr. Vassilis Liakos, for his information on the traditional beekeeping in the Grevena area; to the senior researcher of the Hellenic Folklore Research Centre of the Academy of Athens, Dr. Andromachi Economou, for her help with some Arvanitic words (i.e. words from an old-Albanian dialect spoken in Greece) concerning the beekeeping of Attica; to the researcher of the Hellenic Agricultural Organization "Dimitra", Dr. Fani Hatjina for permission to use photographs; to the educator, Dr. Panagiotis Veltanisian, for permission to use a photograph from his archive concerning the wicker hives of Salamis;

to the environmental cartographer, Dr. Georgios Tataris, for permission to use his own created maps; to the researcher, George Speis for permission to use photographs; to the publisher and the editor of the journal *Melissokomiki Epitheorisi* (Beekeeping Review), Irene and Filippos Pappas respectively, for permission to use several photographs published in this journal; to the publisher of the journal *Arnaia*, Dimitrios Kyrou, for permission to use photographs related to the beekeeping of Arnaia (Chalkidiki); to the agronomist, Vardis Sellianakis, for permission to use images from his unpublished undergraduate thesis at the Agricultural University of Athens; and to the beekeeper, Ioannis Protopsaltis, for sending me photographs from his personal beekeeping collection. Last but not least, I would like to thank my informants, old beekeepers from all over Greece.

When, some twenty years ago, I began engaging with issues related to archaeology, ethnography and history of beekeeping, modeled on the work of the late Eva Crane (1912-2007), I could not imagine that she would "contribute" (in a metaphysical sense), through the Eva Crane Trust, to the publication of my study. I therefore wish to express my deepest gratitude to the Eva Crane Trust, both for its financial assistance, without which it would have been impossible to publish this book, and for the trust shown in me. Both are a real honor to me.

INTRODUCTION

Basketry is an ancient art, older than pottery; however, it is unknown to us when man used woven hives for the first time to practice beekeeping. The oldest relevant reference dates back to only the 1st century BC and comes from the Latin writer, Marcus Terentius Varro (116-27 BC), who, among other things, examined the materials used for the construction of hives in his time. However, it should be considered almost certain that the ancient Greeks also used wicker hives and simply no information of their use exists in the surviving literature, just as it is not found in general about the hives used in Greece at the time.

The first to talk about woven hives in Greek literature were Saints John Chrysostom and Gregory of Nyssa in the 4th century AD. Hesychius of Alexandria, however, was the one who in the next century explains the entry "hive" in his *Lexicon* as a "woven bee vessel" (πλεκτὸν ἀγγεῖον μελισσῶν), which allows us to assume a widespread use of wicker hives in his time.

The Greek area was, from antiquity until the establishment of modern beekeeping in the 20th century, that intermediate zone where both horizontal types of hives (typical of Africa and the Near East) and upright ones coexisted. The latter, in their open-at-the-bottom form, constituted – with a few exceptions – the pre-eminent type of hive in the Balkans and the Central and Western Europe. In their open-at-the-top with top-bar form, however, they were used, over time, exclusively by beekeepers in Greece and especially in its southern part.

All the main types of traditional hives of the Greek beekeepers, namely the horizontal, open-at-one-end or open-at-both-ends, as well as the upright, open-at-the-bottom or open-at-the-top (which in some cases were open-at-the-top-and-bottom), were also used in their woven form. Almost all types of wicker hives had sub-types, which arose mainly from the way beekeeping was practiced in different regions. Thus, a particularly large number of types and sub-types of wicker hives appears in Greece, the like of which are not found elsewhere.

In some areas, wicker hives were the only type of traditional hive in use; in others it was the most used, while there were also areas where woven hives were sporadically found or were completely unknown (mainly on islands).

The advantages of wicker hives over hives made of other materials were several and important for their beekeeper user. In the first place, one could make them relatively easily on one's own with materials found in nature. Of course, there was the possibility to obtain them from a professional basket weaver, but this was not necessary, as for example in the case of clay hives, which had to be bought from a potter. But even in cases where one would buy the wicker hives from a basket weaver, these were normally cheaper than the corresponding clay hives. Another advantage was that the construction of wicker hives did not require painstaking work on the part of the beekeeper, as was the case with hives made of tree trunks, with built hives, and with hives constructed out of a piece of porous rock.

The wicker hives, in their final form, that is, smeared with a mixture of mud and dung and with the protective materials that were placed on them for additional protection (straw hackle, stone slab, etc.) provided very good insulation to the bee colony, both in the heat of summer, and in the cold of winter. The material also allowed the bee colony to "breathe", so that moisture does not remain in the hive. Finally, the wicker hives were the ideal hive for migratory beekeeping, because they were light, which greatly facilitated the beekeeper in loading and unloading, and reduced the number of required routes through contemporary transport, allowing the transportation of more hives per route. They were also not in danger of breaking during the stresses of handling during transportation, as was the case with clay hives. Their only drawback was their relatively low durability over time.

This is the first attempt to examine the various types and sub-types of Greek wicker hives, both in relation to the methods of practicing beekeeping and to their construction. It was also considered important to investigate their use by Greek beekeepers over time, as well as the influence that some of them exerted on the evolution of world beekeeping.

The research, apart of course from the existing modern literature, ethnographic and beekeeping-related, was based on four pillars: a) the study of the relevant ancient and Byzantine literature; b) the study of the works by foreign travelers who visited Greece from the 17th to the 19th centuries to identify elements related to Greek beekeeping and wicker hives; c) the study of archival, mainly unpublished, material; and d) the on-the-spot investigations carried out by the author, between the years 2006 and 2022, almost throughout the country.

I

WICKER-HIVE TYPES AND THE PRACTICE OF BEEKEEPING

1.1 Upright Wicker Hives

1.1.1   Open-at-the-top and open-at-the-top-and-bottom movable-comb wicker hives

Movable-comb top-bar wicker hives have the shape of an inverted truncated cone. The diameter of the mouth in their upper part is larger than that of the base, and they possess oblique, inward-converging walls from top to bottom. On the mouth, is placed as a kind of roof, a series of tangential bars to which the bees attach, one by one to each bar, their honeycombs. In this way, honeycombs are created which hang attached only to the said bars and can be moved, allowing for a series of beekeeping manipulations that are impossible to carry out in fixed-comb hives. The principle of operation of the movable-comb hive is the same as that of the frame hive used in modern beekeeping. Besides the latter in both forms, that of the developed and that of the developing world, derives its origin, as we will see in Chapter 4, from the traditional movable-comb wicker hives of Greece.

Top-bar wicker hives and generally top-bar hives, regardless of their manufacturing material, were in use in recent centuries and until a few decades ago in some areas of southern Greece. These areas form on the map an arc, one end of which is located in central-western Crete, its curve embraces Cythera, the eastern Peloponnese and the islands of the Argo-Saronic Gulf, and its other end rests on Attica and the island of Kea (FIG. 1). Apart from this arc, nowhere in the world, in the regions of the so-called western bee (*Apis mellifera*), that is, in Europe, Africa, and the Near East to Iran, were traditional top-bar hives ever used.

As for the individual characteristics of the top-bar wicker hives, the obliquely converging walls towards the bottom are a prerequisite to prevent the bee's tendency to attach its combs to the side walls of the hive. When the walls are vertical or oblique in the opposite direction, that is, converging towards the top of the hive, the bees attach their combs on them as well. When, however, the slope of the walls is inwards, then the insects attach the combs only from the roof of the hive.

FIG. 1. Areas where movable-comb hives were used. A red border encloses the areas where movable-comb hives were exclusively or mainly used; a green one, the areas where movable-comb hives were used in parallel with fixed-comb ones.

Top-bar wicker hives were of two types: the open-at-the-top (FIG. 2), and the open-at-the-top-and-bottom (FIG. 3). In some cases the base of the former was not flat, but protruding downwards (FIG. 4). The hive was then resting on a few stones placed on the periphery of its protruded base.

The width of the top-bars was extremely important in these wicker hives because it was on this that the creation of movable-combs depended. It would have to be about

3.5 cm in order to have the so-called "bee space"[1], that is, the necessary distance between two honeycombs. This space was empirically created by the traditional users of top-bar hives according to the width of the top bars, which they calculated roughly by using as a measure the length of the big phalanx[2] of the thumb or the sum of the width of the forefinger and the middle finger[3].

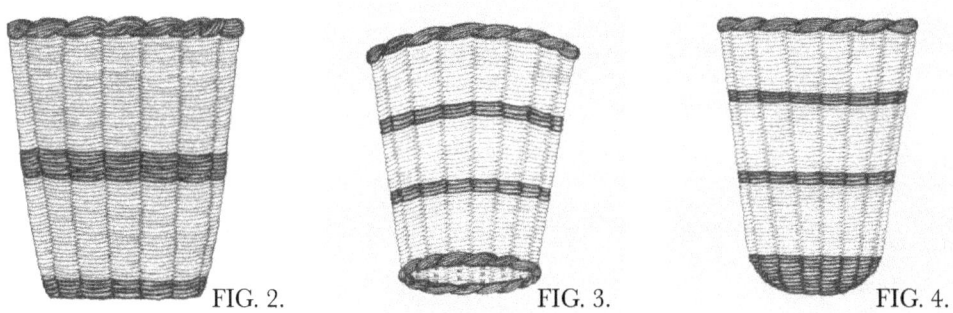

FIG. 2. Open-at-the-top wicker hive
(Drawing: V. Chatzilakou, from Efthymiou-Chatzilakou 1979/80).

FIG. 3. Open-at-the-top-and-bottom wicker hive
(Drawing: V. Chatzilakou, from Efthymiou-Chatzilakou 1979/80).

FIG. 4. Open-at-the-top wicker hive with downwards protruding base
(Drawing: V. Chatzilakou, from Efthymiou-Chatzilakou 1979/80).

The existence of movable combs in these hive types allowed easy and unhindered control of the inside of the hive for a number of reasons, such as the early diagnosis of diseases and other problems coming from the enemies of the bee colony. It also allowed the immediate inspection of the condition of the bee colony regarding the satisfactory population, the existence of enough food, the existence of queen cells, etc. Movable combs also allowed the easy colony multiplication, without hunting and capturing of swarms, the prevention of swarming and, of course, easy harvesting[4]. However, the advantage of movable combs was at the same time the main disadvantage of these hives when they were transported. This is because their honeycombs, as they were attached only from the roof, were easier to cut off and fall into the hive, with all that this entails for the bee colony. There were, however, in many areas, beekeepers who, with appropriate loading and stacking methods and due care, used movable-comb hives for migratory beekeeping as well[5].

The earliest reference to open-at-the-top wicker hives is attributed to the French traveler, physician and antiquarian, Jacob Spon (1647-1685), who together with the

Englishman, George Wheler (1650-1723), visited, in 1676, the Monastery of Kaisariani in Attica, where such hives were in use by the monks. Two years later, Spon included in his travelogue a description of the movable-comb hives of Hymettus. According to his account "their hives are covered with five or six small boards, where the bees build their honeycombs, with a small straw roof on top. So when they want to separate their hives, they only have to remove half of the boards that carry the adhering honeycombs and place them in a new hive. In order for the bees to be less frightened, they expect a part of them to be in the countryside, and then they put a new hive in place of the old one; so when they [i.e. the bees] return at night, they think it is their old home and finding nothing inside they start building their cells"[6].

In the title of his work, Spon also added the name of his companion, Wheler, to honor him. However, when the latter learned that Spon's work was to be translated into English, thinking that the participation in the risks and expenses of a travel entails the commonality of scientific results, he decided to translate and publish it under his own name[7]. Since then, much ink has been spilled on the issue of Spon's plagiarism by Wheler. However, to be fair, it should be noted that with regard to movable-comb wicker hives that are of interest to us here, Wheler's book refers more extensively and in more detail to them, even citing a relevant drawing (FIG. 5)[8].

FIG. 4. Open-at-the-top wicker hive according to George Wheler (1682, 412).

So according to Wheler: "The hives they [i.e. the monks of Kaisariani Monastery] keep their Bees in, are made of Willows, or Osiers, fashioned like ours common Dust-Baskets, wide at the Top, and narrow at the Bottom; and plaistered with Clay, or Loam, within and without. They are set the wide and upwards, as you see here, (A.B.) The Tops being covered with broad flat Sticks, (as at C. C. C.) are also plaistered with Clay on the Top; and to secure them from the Weather, they cover them with s Tuft of Straw, as we do. Along each of those Sticks, the Bees fasten their Combs; so that a Comb may be taken out whole, without the least bruising, and with the greatest ease imaginable. To increase them in Spring-time, that is, in *March* or *April*, until the beginning of *May*, they divide them; first separating the Sticks, on which the Combs and Bees are fastened, from one another with a Knife: so taking out the first Combs and Bees together, on each side, they put them into another basket in the same Order as they were taken out, until they have equally divided them. After this, when they are both again accommodated with Sticks and Plaister, they set the new Basket in the Place of the old one, and the old one in some new Place. And all this they do in the middle of the day, at such time as the greatest part of the Bees are abroad; who at their coming home, without much difficulty, by this means divide themselves equally. This Device hinders them from swarming, and flying away. In *August* they take out their Honey; which they do in the day-time also, while they are abroad; the Bees being thereby, they say, disturbed least. At which time they take out the Combs laden with Honey, as before; that is, beginning at each out-side, and so taking away, until they left only such a quantity of Combs in the middle, as they judge will be sufficient to maintain the Bees in Winter; sweeping those Bees, that are on the Combs they take out, into the Basket again, and again covering it with new Sticks and Plaister"[9].

As can be easily deduced from the comparison of the two texts, Wheler is more interested in beekeeping, and gives clear information. Both travelers, however, refer to the multiplication of colonies without swarm hunting that made an impression on them, since it was unknown in their countries. Spon wrongly speaks of the existence of five to six top bars per hive. A similar number of top bars has not been recorded anywhere in movable-comb hives for the beekeeping with *Apis mellifera*. The top bars were over time at least nine, usually ten to twelve and up to fourteen[10]. In addition, a hive of five or six top-bars would have an extremely narrow mouth (17 to 21 cm), which would create insolvable problems during the practice of beekeeping. So it seems that Spon, who was known to be mainly interested in antiquities, did not pay much attention to wicker hives he encountered in Kaisariani.

It is not clear whether Wheler's reference to the separation of bees ("divide themselves equally") upon their return to the hive, the top bars of which had meanwhile been divided in equal numbers into two hives, is a view espoused by the beekeepers of the monastery or is one of his own. In any case, today we know that all the foraging bees would return to the place where their old hive was before, that is, they would enter the new hive that was placed in the old hive's place. Wheler's description of the hive harvest fully corresponds to reality, as has been recorded many times in the 20th century on movable-comb hives.

Almost seventy years after Spon and Wheler, the top-bar wicker hives of Attica were described, again in the Monastery of Kaisariani, by a traveler named Charles Thompson[11]. His book was a commercial success and witnessed several reissues, but in the process it turned out that its author was an imaginary person[12], possibly some scribbler hired by the editor, who gathered and paraphrased various excerpts, with some skill (to be honest), from texts of other travelers. Regarding the "description" of the wicker hives of Hymettus, this is based, as I have seen from the comparison between the texts[13], on the relevant account of Wheler. However, although the undersigned "Thompson" offers nothing new about the movable wicker-hives, what he describes is not inaccurate (he seems to be particularly attentive to this) and is rendered in a more eloquent way than in its original[14].

In 1790, Abbé Della Rocca of Syros Island (1738 - after 1810), in his important three-volume work on beekeeping, speaks of the top-bar wicker hives of Crete. He had not seen them himself, but as for their use on the island he writes that this "has been confirmed by more than one person"[15]. In fact, it seems that he had fully understood how they worked and refers to them in detail[16]. After all, the hive of wooden boards that he invented and proposed for a more rational practice of beekeeping is based precisely on the method of movable combs.

In 1811, the Scottish novelist, John Galt (1779-1839) arrived, as part of his tour of Greece, in Porto Heli, Argolis. There, walking a little further from the place where he left the ship, he found himself in the midst of more than five-hundred hives that formed, as he says in a literary style, "several cities, towns, and villages, the property of different proprietors"[17]. The hives are described as covered with "earth", as the traveler perceived the mixture of mud and dung that was used for anointing the walls and covering the top bars in the movable-comb wicker hives of the eastern Peloponnese, which were probably the type of hives he encountered[18].

In 1842, William Charles Cotton (1813-1879) published his book, *My bee book*, a compilation of works by earlier authors. There, among other things, he included a woodcut, signed by someone named Whimper, which seems to accompany a text by the traveler, John Sibthorp (1758-1796) on the plants of Attica[19]. The woodcut depicts a Greek beekeeper dressed in *fustanella* (the traditional Greek kilt) having already opened, now examining one of his open-at-the-top hives, outside his house, watched by his wife and child (FIG. 6). For this woodcut there is no information and despite my intense efforts I did not manage to find any evidence[20]. However, as everything suggests[21], the hives depicted cannot be other than the movable-comb wicker ones[22].

FIG. 6. A 19th-century woodcut.
A beekeeper inspects an open-at-the-top wicker hive (Cotton 1842, 107a).

Cotton's book also contains information about the beekeeping of Attica gathered by the traveler John Hawkins (1758-1841)[23]. Hawkins's account is divided into three parts: the first two parts include information from two informants: the abbot of the Monastery of Penteli (or Menteli as it was then known) and an 80-year-old Athenian beekeeper, named Bueras, while in the third there appear some notes by Hawkins himself on the account of the beekeeper.

In the last decade of the 18th century, that is, the period when Hawkins was in Athens, the abbot of the Monastery of Penteli, according to Dimitrios Kambouroglou[24], was Kyrillos Degleris, who is supposed to be the informant of the English traveler[25]. The abbot in question, although he is obviously talking about movable-comb wicker hives, does not describe them. At some point, however, he mentions that the stabilization of each hive was achieved by digging a shallow hole at the spot of its placement and using four stones to fix it therein[26]. This description refers to an open-at-the-top wicker hive with downward protruded base.

Of all the interesting information taken from the abbot, it is worthwhile dwelling on the south orientation of the hives by the monks, so that the bees could be protected from the northern winds; on the destruction of the queen cells with their hands or the killing of the queens when they left with the swarms so as not to weaken the bee colonies;

on the change of the position of the plundering hives with that of the plundered; and on the non-cleaning of the hives by the monks, because of their large number, in contrast to some villagers who cleaned them having only a few hives to manage[27].

The monks of the Monastery of Penteli also knew that the bee colonies do not tolerate a queen from another hive, which they inevitably kill. In addition, they accept combs with brood from another bee colony, from which they produce a new queen. As for their enemies, the worst one was considered to be the bee-eaters (*Merops apiaster*), and the measures they took against them were to destroy their nests, which were built in rock caveats and ravines. Another enemy were badgers (*Meles meles*) who overturned the wicker hives and ate the honey. Wasps (*Vespula* spp.)[28] were not ranked by the abbot among the most important enemies. In his opinion, these attacked only weak bee colonies, being easily persecuted by the healthy and strong ones[29].

Here it should be mentioned that in the Monastery of Penteli, which was the most populous and one of the richest in Attica[30], beekeeping was practiced intensively. The French consul in Thessaloniki at that time, Felix Beaujour, mentions in his report from Athens, which he had visited, that Penteli had 1,200 hives, while another 3,000 were active in the four monasteries of Hymettus, and 2,000 were in various monastic estates; a total of 6,200. The villagers of Attica had as much in their possession[31]. The vast majority of these hives consisted of open-at-the-top wicker hives and their beekeeping seems to have been exclusively static[32].

The second part of Hawkins' text is tellingly entitled *Answers from Buera of Athens, aged eighty years*, and refers to the information given by the beekeeper Bueras to the English traveler about the beekeeping in Athens. But who was this Bueras, whose beekeeping knowledge, as we will see below, is surprising even today? His surname[33] points to him being of Arvanitic origin (Albanian speaking Christian)[34], which may possibly explain the use of slightly different form of Greek words, such as *thymasi* instead of *thymari* (thyme) and *kosinia* instead of *kofinia* (wicker hives), found in the text[35].

Be that as it may, Bueras describes the wicker hive of Athens, which had a downward protruded base, and of which Hawkins cites a drawing with its dimensions in inches (FIG. 7). According to that, its height amounted to 17 inches, that is, 43 cm, while the diameter of the mouth reached 19 inches, that is, 48 cm. When the year was good for the bees, according to Bueras, it was deemed "necessary to enlarge the *kosinia* [wicker hives] two fingers by extending the wicker work"[36]. The wicker hive was internally whitewashed, while externally it was anointed with a mixture of dung and mud. Its entrance was two inches (5.08 cm) long, one inch (2.54 cm) tall, and was four inches (10.16 cm) from the base. The number of top bars ranged from twelve to fourteen. Above the top bars was placed a bunch of plane-tree sprigs along with their leaves, and on top of them a "sheaf of straw". The number of wicker hives belonging to Athens, in the mid-1790s, amounted to some 3,400[37].

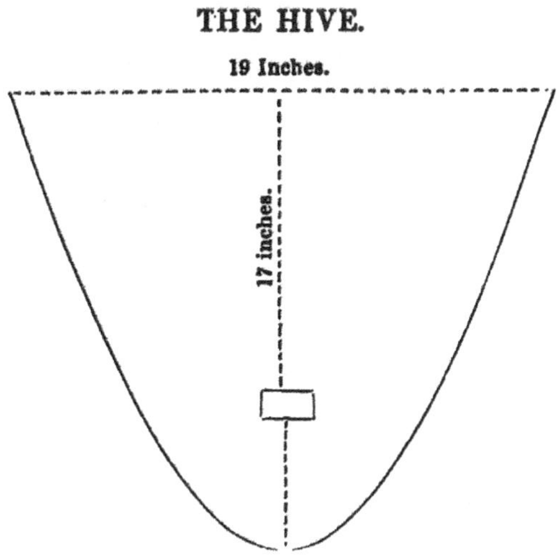

THE HIVE.

19 Inches.

17 inches.

FIG. 7. Rough drawing of an open-at-the-top wicker hive with downwards protruding base by John Hawkins (Cotton 1842, 105b).

There is no doubt that Bueras was ahead of his time in terms of beekeeping knowledge, as well as of some of his practices. He was thus able to make the moſt of the possibilities offered by the movable-comb hive. He knew, firſt of all, the sex of the queen and the fact that she is the mother of everyone in the hive. This was not at all a given in his time for a beekeeper, especially in Greece that was under Turkish occupation. He knew the time required to hatch a queen[38], as well as that a new queen cannot be directly introduced into a queenless hive, because the bees themselves will kill her. This is true, but there are methods, known nowadays, that allow the introduction of a queen into a queenless hive, something that Boueras does not seem to have known. However, he had a way of overcoming this weakness by placing in the queenless hive a comb with brood from another hive, from which the bees created a new queen[39]. Although he considered that it was not possible to unite two weak bee colonies, because they would exterminate each other[40], nevertheless he dealt with it satisfactorily by changing the positions of the hives of the weak bee colonies with those of the ſtrong ones, so that they all acquire about the same bee population[41].

He considered as enemies the bee moth[42], which he refers to as the "worm", the bee louse[43] and the wasps[44]. For the latter, he did not take any measures, believing that it was sufficient to reduce the entry of hives from the insects themselves, with propolis, during the time the wasps attack, and to maintain, as far as possible, ſtrong hives. For the treatment of the bee moth he used vinegar, while for the louse he employed smoke[45].

Regarding his beekeeping practices, worth mentioning is the deſtruction of the queen cells so as not to weaken the bee colony with swarming. If, however, the bee colony finally swarmed, then the queen of the swarm was arrested and killed and the swarm returned to its hive. The multiplication of colonies was carried out by placing in an empty wicker hive four or half of the combs that had brood from a strong hive and exchanging their positions. In the event that a swarm was arrested, it was thrown into a wicker hive in which two honeycombs from another hive were placed. A hive could multiply in favorable years two or even three times which are the most amenable. From the new hives, however, others could be produced in the same year. A beekeeper was therefore satisfied if during the year he doubled the number of his hives[46].

According to Bueras, the most important flowering was that of thyme (*Thymbra capitata*), which in good years bloomed from May 10 (according to the old calendar), and its flowering lasted until August. The harvest was carried out between August 7 and 15, in order to avoid, as they believed, the ominous "drums" (*dies infausti*), that is, the first six days of August. During the harvest, the four middle combs of the hives were always left in the bee colony. Honey from the autumnal heath (*Erica manipuliflora*) was not considered good and, in cases where it was produced, it was sold at half the price of August honey, which consisted mainly of thyme. Heather honey, however, was usually not harvested and was left in the hives[47].

The third part of Hawkins' text includes some of his notes on Bueras' exposition. The traveler seems to have identified the differences of the "Grecian", as he writes, hive to the hives of his homeland and had realized, at least to some extent, the potential of movable-comb hives. However, he distrusts some of his informant's words, even going so far as to write that the method of changing positions between strong and weak beehives, to which Bueras had repeatedly referred, is "perfectly erroneous"[48]. We certainly know today the correctness of the method[49], which is also applied to the modern practice of beekeeping with frame hives.

In 1882, the American author, Frank Benton, a prominent figure of beekeeping and regular collaborator of many beekeeping journals, arrived in Athens. During his stay in the country, he witnessed on Mount Hymettus the use of open-at-the-top wicker hives, which intrigued him. He then bought some of them and practiced beekeeping to see how they worked. A decade later, arguing against the opinion of some German researchers, who claimed that the movable-comb hive was first invented by Johann Dzierzon, he said that the beekeepers from whom he procured the wicker hives on Hymettus had been following the beekeeping system with movable combs for many centuries and had of course no idea about the achievements of beekeeping in Europe and America to be influenced, since they were illiterate[50].

Movable-comb wicker hives were used in the 20th century in various regions, always within the geographic "arc" mentioned above (see FIG. 1). The southernmost area was Crete and mainly its western part, the prefecture of Chania. There, the wicker hives (FIG. 8, 9) were, almost in their entirety, open-at-the-top-and-bottom[51]. Their dimensions were about 38 cm height, 42 cm the diameter of the mouth and 32 cm that of the open base[52]. Apart from the prefecture of Chania, the use of wicker hives has been recorded in some areas of the prefecture of Rethymno, such as in Arkounta[53], where it seems that they were used, in their open-at-the-top form, and in some other mountainous areas of central Crete[54].

FIG. 8. Open-at-the-top-and-bottom and open-at-the-top wicker hives of Crete in the horizontal position (Photo: V. Sellianakis, 1998).

FIG. 9. Honeycomb inspection in a movable-comb wicker hive by a traditional Cretan beekeeper (Photo: P. Papadopoulo, 1938).

These wicker hives were usually placed, like the other traditional hives of Crete, within enclosed stone structures (FIG. 10) called *melissokipi* or bee gardens[55]. Besides, Crete had a tradition in these constructions, which are very likely to have been used on the island since antiquity[56]. Within the bee gardens, each hive had its own stone base, which usually consisted of a stone slab. These bases were sought to have a slight slope so as not to retain water and not to allow the growth of grasses[57]. In the places where the wicker hive rested on its base, mud or a mixture of mud and dung was placed around it, except for about ten millimeters, which was the entrance of the bees[58].

FIG. 10. Bee enclosure with open-at-the-top-and-bottom wicker hives in Kampani, Akrotiri, Chania, Crete in 1939 (Photo: P. Papadopoulo).

Top bars were known in Crete as *kanonia*[59] or *kantineles*[60] and their lower side was often not flat but curved outwards (FIG. 11) in order to make it easier for the bees to attach their honeycombs. After their placement at the mouth of the hive they were anointed on top with a mixture of mud and dung[61]. Above the top bars, for insulation and protection from the elements of nature, small twigs and leaves were placed, and upon them a stone slab or stones or even a clay lid were added[62]. In the famous Gorge of Samariá, the local beekeepers, to protect their wicker hives, instead of twigs, placed pine bark that was more resistant to the ravages of time. In fact, traces of the incisions for the removal of bark are still visible in several trees of the area[63].

FIG. 11. Top-bar of movable-comb hives of Crete (Photo: V. Sellianakis, 1998).

In western Crete, the vast majority of beekeepers practiced migratory beekeeping by moving their wicker hives. This practice was first recorded at the beginning of the 20th century, but it is probably considerably older. The hives were mainly transported from areas with thyme to areas with autumnal heath and vice versa[64]. Other flowering exploited by the beekeepers of western Crete, by transporting their wicker hives, were those of Cretan mountain tea (*Sideritis syriaca*), locally known as *malotira* and chestnut (*Castanea sativa*)[65].

Wicker hives from the area of Akrotiri, after being harvested for the early local thyme honey, were transported by some beekeepers to areas where the *malotira* bloomed. These areas are located in the White Mountains, around the villages of Pemonia and Ramni, but also at higher points, such as Gournes. Then, the hives moved to areas with plenty of autumn heath, usually around the villages of Prasses, New and Old Roumata, where other beekeepers transported hives directly from Akrotiri[66].

Beekeepers in the area west of the White Mountains took advantage in the spring of the flowering of the tree heath (*Erica arborea*), which was found in the same places as its autumn counterpart, and then transferred their wicker hives to areas with thyme, especially around Katochori. Some, however, before transferring to thyme, first moved their hives to the chestnut trees in the village of Topólia. With the completion of the thyme flowering, the hives were returned for the autumn heath and for overwintering (MAP 1)[67].

Transport was carried out by pack animals or, more rarely, by carts. On pack animals, a total of seven wicker hives were loaded, entirely covered with perforated flax cloth. Four hives were placed, in pairs, on each side of the saddle, and two others on top of them in a horizontal position. The seventh hive was placed, also horizontally, in the middle of the saddle. In the horizontally placed hives, attention was paid to the axes of the top bars of their honeycombs to be in a perpendicular position to the ground so as not to cut off the honeycombs along the way. The distance to be covered was usually 15-20 kilometers corresponding to 4-5 hour time. Transfers took place during the night, usually after midnight[68]. In some cases involving the exploitation of Cretan mountain tea, the final part of the transport, due to the nature of the soil, was undertaken by the beekeeper himself, carrying two wicker hives, one on each shoulder, tied together with a rope[69].

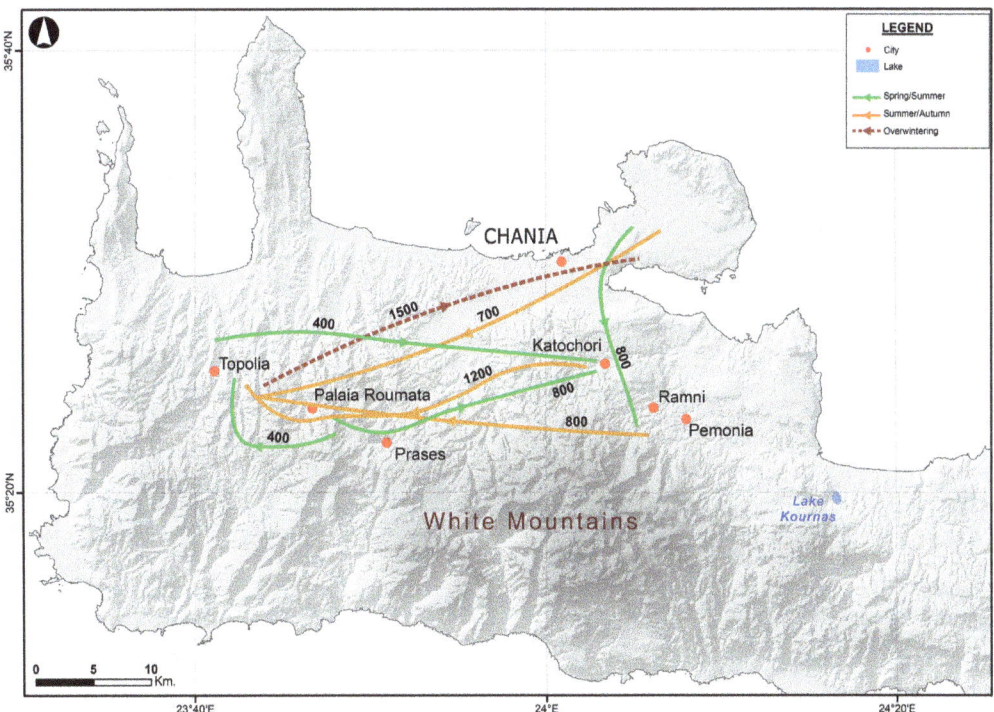

MAP 1. Map showing the migration of hives by the beekeepers of western Crete with their open-at-the-top-and-bottom wicker hives until about the middle of the 20th century (estimated number of hives per destination). (Created by G. Tataris).

As for the traditional tools employed by Cretan beekeepers, using top-bar wicker hives, the most important were the following two: a clay smoker and a metal hive tool. The clay smoker had a double opening, a large one from where the beekeeper blew, and a smaller tubular one, through which the smoke was directed to the hive (see FIG. 8)[70]. Dry horse manure or cow dung was used as fuel[71]. The metal hive tool, necessary for the beekeeper's work, existed in various forms, but its main characteristic was the presence of a fork at one of its end (see FIG. 8). Its length was noticeably longer than modern hive tools, reaching 40-50 cm[72].

A mask made of convex parallel wires (see FIG. 8)[73] or semi-transparent linen cloth[74] was also used. The Cretan bee (*Apis mellifera adami*), which unfortunately seems to have now been disappeared – due to its intermixture with other subspecies – was considered "extremely aggressive" and therefore the use of masks by local beekeepers was almost imperative. The type of traditional wire mask of Crete was found throughout the Aegean. The oldest depiction and description of its construction is due to the Abbé Della Rocca[75]. He considers it superior to those used in his contemporary France, and cites the interesting information that this type of mask was then manufactured, at the end of the 18th century, in large numbers in the industries of Smyrna (nowadays Izmir) that covered the demand throughout the Archipelago[76].

FIG. 12. Open-at-the-top wicker hive of Cythera on its porous-stone base
(Photo: I. Protopsaltis, 2019).

On the island of Cythera, the most widespread traditional hives were the top-bar wicker hives, which were represented by both of their main types, the open-at-the-top (FIG. 12, 13) and the open-at-the-top-and-bottom (FIG. 14, 15). The former, however, were in a marked majority. The diameter of their base was 23 by 25 cm, while the height amounted to 46-50 cm. The top bars were made of wild olive-wood or from the staves of old barrels. The corners of their underside were cut off to delineate the point from where the bees would start the attachment of each honeycomb. On their upper side they were covered with a mixture of dung and clay so that the wicker hive was at the top tightly shut. From above a layer of oleander branches (*Nerium oleander*) was placed, and on this layer a new one made of plane leaves was placed in the past, that was later replaced by sheet metal[77]. The reason for the placement of the layer of oleander branches on Cythera was twofold. On the one hand, to act as insulation for greater protection from the high temperatures of summer and generally from the elements of nature, and on the other hand, to remove scorpions, which had been known to avoid oleander, so as not to threaten the beekeeper. Above all, on these layers rested a schist slab[78].

FIG. 13. Interior of an open-at-the-top wicker hive of Cythera (Photo: I. Protopsaltis, 2019).

FIG. 14. Open-at-the-top-and-bottom wicker hive of Cythera on its base (Photo: I. Protopsaltis, 2019).

FIG. 15. Open-at-the-top-and-bottom wicker hive of Cythera as seen from above (Photo: I. Protopsaltis, 2019).

The whicker hive was usually placed on a stone base that was round and was made of porous stone. However, in the areas where there was no such rock, stone slabs were used from those located around. The bases helped to avoid soil moisture, which affected both the bee colony and the longevity of the hive. The open-at-the-top-and-bottom wicker hives, some of which had an oval shape, clung with the mentioned mixture of mud and dung to the bases, so as not to be swept away by the wind[79]. According to my informant, Ioannis Protopsaltis, the open-at-the-top-and-bottom wicker hives were made on Cythera for greater protection from the thieves. This is because someone who would pick up the hive would be faced with its population[80].

In many cases, such as in the area of Makrea Skala, the wicker hives were placed in bee boles (FIG. 16), in the dry-stone walls that shaped the terraces of the island. The dimensions of these bee boles, without being fixed, were no less than 50 cm in width and depth, and 60 cm in height[81].

FIG. 16. Bee boles in dry stone walls of agricultural terraces, Cythera Island
(Photo: I. Protopsaltis, 2022).

In order to deal with hornets (*Vespa orientalis*), which were and still are a major problem for beekeeping, especially on the islands, the beekeepers of Cythera used reeds to create a new entrance, which they adapted to the entrance of their wicker hives[82]. From this new entrance, which was placed in the hives at the time of the hornets' attacks, the bees passed through, but the large hornets could not do so. With the narrow entrance the bee colony was also better defended against wasps (*Vespula* spp.).

The tools used by the beekeepers of Cythera were similar to those of their Cretan colleagues, namely the metal hive tool with its one end in the shape of fork, the mask made of wires and the clay smoker with a small and a larger opening[83].

Harvesting was carried out once a year, after thyme flowering, and during this process honey was received only from the combs on one side of the wicker hive. The following year, the honeycombs of the other side were taken in the harvest. The central combs, being four at least, where the brood was located, were of course left unscathed. Production was small and a steady yield of around 2.5 kg per hive was considered satisfactory[84]. However, the thyme honey of Cythera enjoyed, and still enjoys, a great reputation, and is considered of top quality.

In the eastern Peloponnese the only type of traditional hive was the top-bar wicker one. The latter, however, were also found, although rarely, in the southern part of the Peloponnese, in some areas of Laconian Mani and Messenia, but in parallel with other types of hives.

Regarding their use in Messenia[85], we only know that the wicker hives were open-at-the-top[86]. In the Laconian Mani they were used at the end of the 19th century in the village of Koulouka, near Gythio[87].

The area par excellence where the top-bar wicker hives were found in the Peloponnese was the eastern part of the prefecture of Laconia, known as Kynouria, the prefecture of Argolis, Troizinia, and the eastern part of the prefecture of Corinth. In the southern part of this area the wicker hives were open-at-the-top-and-bottom, while in the north part two sub-types of open-at-the-top hives were used.

The open-at-the-top-and-bottom ones have been recorded in Laconia[88], in the area of Mount Parnon, and in Kynouria[89]. In Pistámata, Laconia and generally in the area of Zárakas, these hives were formerly covered with densely leafed branches of various trees or with straw hackles. In modern times they were covered with tin or stone slabs. The wicker hives were placed on a stone slab, in a leeward and sunny place, within a fenced area. The latter served to protect mainly from stray animals, which, under certain conditions, were able to cause damage to the hives[90].

The harvest was carried out by two people in August. A total of six honeycombs were harvested, three on each side of the wicker hive. The local beekeepers seem not to have used a clay smoker, but simply, during the harvest, one of the two participants, lit, somewhere casually, a piece of dry dung and blew the smoke towards the hive. The metal tools that were used in the harvest were two, a knife and a hive tool[91].

The hornet (*V. orientalis*), was considered in the region the greatest enemy of bees, and beekeepers took various measures against it. They made special traps with baits or killed the insects in question themselves, beating them with branches in the apiary, especially at noon, when they went in front of the entrances of the hives trying to grab bees. The bee-eaters, who pass through the area twice, April to May and August to

September, were confronted by shooting in the air for intimidation; more rarely they were killed[92].

In Daimonia, Laconia (FIG. 17) bars were not used on the hives. Instead, cut and peeled branches, certainly of circular cross-section were used. In Mount Parnon area, the wicker hives (FIG. 18) often had a small capacity, around twenty liters, while in some cases, as in a wicker hive from the museum collection of the Institute of Agricultural Sciences, the internal volume did not exceed fifteen liters[93].

FIG. 17. Open-at-the-top-and-bottom wicker hive from Daimonia, Laconia
(Photo: G. Mavrofridis, 2007).

FIG. 18. Open-at-the-top-and-bottom wicker hive of extremely small capacity from Mount Parnon, Laconia (Photo: G. Mavrofridis, 2007).

These open-at-the-top-and-bottom wicker hives were also used for migratory beekeeping, except that here the transport, in contrast to Crete, was mainly carried out by sea. According to Angelos Typaldos-Xydias, the beekeepers of Kynouria and especially of Leonidio transported their hives with small sailing vessels to the island of Spetses and to the opposite coasts of Argolis[94], probably for the flowering of thyme (MAP 2) In fact, it seems that a similar transport was witnessed by the traveler John Galt in 1811, as mentioned above.

In Argolis, Troizinia and the east of Corinth Prefecture the wicker hives were open-at-the-top. There were in parallel use two sub-types: one with a flat base and the other with a downward protruding base. The second sub-type allowed the wicker hive not to touch the ground (or on some stone slab) like the other open-at-the-top ones, but to set on stones placed around the periphery of its base, so that it would not rot quickly (FIG. 19).

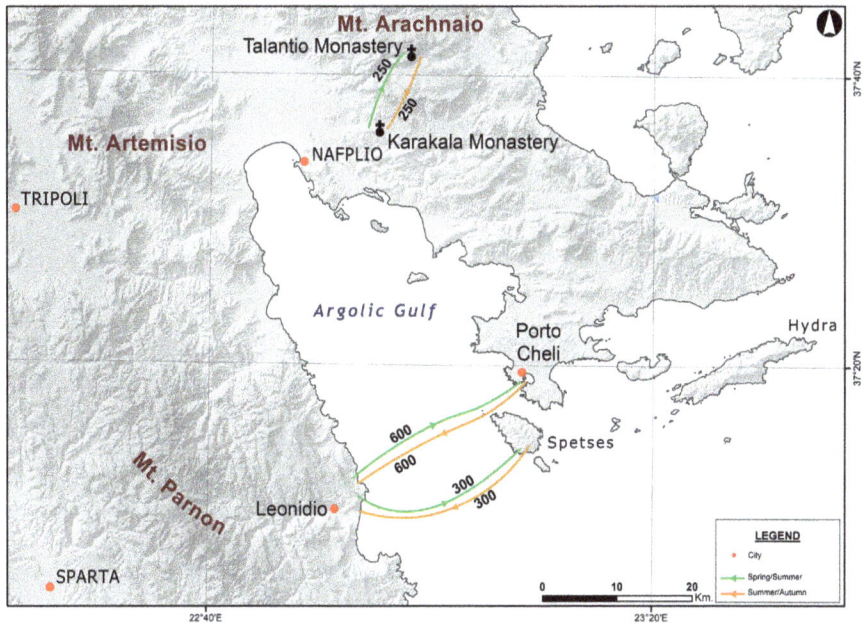

MAP 2. Map showing the migration of hives a) by the beekeepers of Kynouria by sea and b) by the girls of Arachnaio village on behalf of the Monastery of Talantio (the hives used were movable-comb wicker ones, open-at-the-top-and-bottom in the first case, and open-at-the-top in the second; estimated number of hives per destination - created by G. Tataris).

The top-bars were flat on their upper side, but on the lower side they had a triangular shape or the edges chipped, so that it would be easier for the bees to build their combs one on each bar. All of these bars were covered from above with a mixture of mud and dung[95]. Above the bars were placed, crossed, branches of plane trees or sprigs of plane trees that grew by the streams. Then a straw hackle covered the wicker hive, protecting it from the rain and the summer sun. The wicker hives rested on some stones to avoid moisture[96].

In addition to individuals, top-bar wicker hives were also owned by the monasteries of the area. The Monastery of Agios Demetrios Karakala, Argolis was in possession of about a hundred wicker hives. A dependency (metochion) of Karakala Monastery, named Talantio Monastery, in the middle of the last century, cultivated two to three hundred wicker hives. These hives were transferred every year from the Monastery of Tallantio to that of Karakala for wintering, due to the milder climate that prevailed there and the abundance of autumn heath. The transfer was undertaken by the girls of the village of Arachneo, who each carried a hive the relatively long distance between the two monasteries. The return of the hives to Talantio Monastery was implemented in March in the same way (MAP 2)[97].

25

FIG. 19. Open-at-the-top wicker hive with downwards protruded base from Karatzas, Troizinia (Photo: G. Mavrofridis, 2010).

In the area of Arachneo, transfers were also carried out by individuals, but at short distances, usually up to 4-5 kilometers, within the mountainous plateau in which the village is built[98]. In the other areas of the northeastern Peloponnese, these top-bar wicker hives show that an exclusively static beekeeping was practiced. In any case, the hives were harvested once a year, usually in the first part of August.

In some cases, open-top wicker hives were placed for protection inside recesses in stone walls. One such wall, which had 98 bee boles for open-at-the-top wicker hives, was recorded and photographed in 1952, east of Mycenae by Brother Adam[99]. Eva Crane, who sought the construction in the area in the late 1960s, says she was informed that the wall with the bee boles had been demolished to build a factory in its place[100]. However, years later, Thanasis Bikos located this wall in the village of Monastiraki in Argolis[101], about one kilometer southeast of Mycenae. In July 2022, I carried out a fieldwork in the area, as part of a more general research by the Laboratory of Biogeography and Ecology of the Department of Geography, University of the Aegean, on the stone-built beekeeping constructions of Greece. The research revealed interesting information about the use of this wall (FIG. 20) and made known the year of its construction in 1905[102].

FIG. 20. Wall with bee boles for open-at-the-top wicker hives in Monastiraki, Argolis (Photo: G. Mavrofridis, 2022).

The beekeepers of Troizinia protected their apiaries by placing them in naturally guarded areas or in a fence created of cut branches. The need to protect the hives in Troizinia arose from the existence of many herds of animals, which created problems to apiaries, when the latter were not fenced[103].

In Karatza, Troizinia, the construction of woven rings, about 20 cm high, has been recorded, which a certain beekeeper, Epaminondas Giakoumis, adapted to the mouth of his wicker hives during the years of rich nectar-gathering. At the mouth of this extension, he placed top bars, while removing an outer bar from the existing ones so that the bees could climb to this second "floor" of the wicker hive. The bees built combs below the top bars of the new space, in which they placed honey. The beekeeper, being one of my informants too, insisted that the use of the extension was his own idea, which he even conceived from descriptions of modern frame hives and the way they worked that he heard from others. Until then he had not seen the hives themselves. The beekeeper Boueras, however, as we saw above, mentions, at the end of the 18th century, an augmentation of the top-bar wicker hives through the extension of the wicker work, in case the year went well for the bees.

In the same village they multiplied their bee colonies, in a slightly different way from the "classic" one, that is, through the division of one wicker hive's combs and bees into two, as already described by Wheler in 1682. They observed their hives, and when they saw queen cells, they transferred their combs to empty wicker hives, destroying at the same time any queen cells that may have existed in the combs that remained in the old hive. In this way, and provided that care was taken to keep the queen in the original wicker hive, the time of acquiring a new queen in the bee colony of the new wicker hive was noticeably reduced. In the harvest they always left, for the needs of the colony, four combs in the middle, out of the 10-12 that each wicker hive had. In the nearby settlement of Tsoukalia, however, the local beekeepers left only one or one-and-a-half combs (they cut from the one the part that had honey, leaving only that with the brood). Interestingly enough, the bees survived and this was due to the fact that there was in the area plenty of autumn heath and strawberry tree (*Arbutus unedo*), and the bees worked well in the autumn, replenishing the harvested honey[104].

On the islands of the Argo-Saronic Gulf, the open-at-the-top wicker hives had a flat base and were the only type of hive used. On Hydra, these wicker hives were about 50 cm high, and were covered for protection with small branches of pine trees, while on Spetses, they placed a straw hackle. The main beekeeping flowering was that of thyme, and harvesting was carried out once a year. During the harvest, four or three of the honeycombs in the middle of the hive were left unharvested[105].

FIG. 21. Open-at-the-top wicker hive from the island of Aegina (Photo: G. Mavrofridis, 2008).

Further north, in the Saronic Gulf, wicker hives were found on all the inhabited islands, except for Agistri where in the past beekeeping was not practiced[106]. Although it is known from informants of mainland Troizinia that the open-at-the-top wicker hives were used on the island of Poros, there is no other information. On Aegina, beekeeping was practiced in the past in a single village, Mesagros, by only two families. The dimensions of the Aeginean wicker hive (FIG. 21), according to the one I had in my possession[107], were 42 cm height, 50 cm mouth diameter, and 36 cm base diameter. The local wicker hive was a little larger in diameter than those used in other areas, and for this reason it usually had 12 to 13 top bars. For protection, the hive was covered with a straw hackle[108].

On the island of Salamis, wicker hives (FIG. 22) were constructed higher than those of Aegina, reaching 55-60 cm, but also narrower, usually receiving 9 to 10 top bars. Top bars on Salamis, however, had their lower side semicircular, in other words they had the shape of the branch cut in half (lengthwise). This local peculiarity was intended to make it easier for bees to attach their honeycombs one by one to each bar[109].

Each wicker hive was covered for protection with a hackle created from *asprohortaro* or white grass (*Andropogon hirtus*), a local grass, and had a height of at least 1.20 meters, so that it could cover the hive while opening (FIG. 23). These hackles had a fairly long life span; they were renewed every ten to fifteen years[110]. In some cases, a metal ring was placed on the cap, with stones inside (FIG. 24), for greater stability when strong winds were blowing[111].

Wicker hives on Salamis were harvested once a year, after the end of the flowering of the thyme. As a rule, on Salamis, wicker hives were harvested only from one side of the brood nest, one year, and from the other the next year, as the beekeepers of Cythera did, so that the bees were left with the necessary food for wintering. The beekeepers of Salamis had developed a special technique to check, without opening the hive and lifting the honeycombs, if the hive had to give honey. They removed the straw hackle and beat the top bars with their finger. Depending on the sound produced, they did or did not proceed to a harvest[112].

In Attica, wicker hives were of the open-at-the-top type (FIG. 25, 26, 27). The information we have about the use of open-at-the-top wicker hives in Attica in the 20th century concerns the Thriasio Plain, the Athenian Plain, the Mountains of Parnitha, Hymettus, and Penteli, and the area of Mesogaia[113].

FIG. 22. Beekeeper, S. Vasileiou lifts a honeycomb from a wicker hive on the island of Salamis in 1956 (Unknown photographer, from Bikos 1995d).

FIG. 23. Demonstration of straw-hackle construction for wicker hives by beekeeper, S. Papanikolaou on Salamis, 1995 (Photo: T. Bikos).

FIG. 24. Apiary of movable-comb wicker hives on Salamis, 1962
(Unknown photographer, P. Veltanisian Archive, Salamis).

FIG. 25. Open-at-the-top wicker hive of Attica with its straw hackle
(Photo: G. Mavrofridis, 2007).

FIG. 26. Apiary of movable-comb wicker hives in Phyli, Attica
(Photo: Eva Crane Trust, maybe 1970s).

FIG. 27. Apiary of movable-comb wicker hives in Pikermi, Attica at the beginning of the 20th century (Unknown photographer, from Toufexis 1909, 97).

In the Athenian Plain and the surrounding mountains, wicker hives had a straw hackle for protection. A peculiarity that was found in some apiaries of the area in the second half of the 20th century concerns the placement, above the hackle, of a metal ring (see FIG. 25, 26) and of stones inside[114], as in the aforementioned apiary of Salamis.

The majority open-at-the-top wicker hives of Attica in the 20th century, in contrast to those recorded in the late 18th century, had a flat base rather than protruding downwards. However, there were also beekeepers who used hives with a protruding base. A photograph of the 1970s depicts a beekeeper of Aspropyrgos (FIG. 28) using this sub-type of open-at-the-top wicker hive[115].

FIG. 28. Beekeeper, G. Nezis with an open-at-the-top wicker hive with downwards protruding base, in Aspropyrgos, Attica, in the 1970s
(Photo: M. Efthymiou, from Efthymiou-Chatzilakou 1980/81).

In the area of Mesogaia, the flat-based open-at-the-top wicker hives rested on four stones. However, the element that differentiated them from the rest of Attica was that it was not a hackle made of straw placed on them, but a thick layer of branches of lentisk (*Pistacia lentiscus*), pressed and artfully tied together, having a stone slab upon. The harvest was carried out here at the end of July and, alternately every year, was harvested only one side of the hive so that a sufficient amount of honey remained for the overwintering of the bee colony[116].

On the neighboring to Attica island of Kea, the main traditional hives used were top-bar ones. Although the island's hive par excellence was of clay, in some cases local beekeepers also used wicker hives. Little information has survived about them, such as their use in a migratory beekeeping exercise, always within the boundaries of the island. The transfer was carried out either by pack animals, upon which two hives were loaded on either side of the saddle, or by the beekeeper himself, who carried a hive on his back[117].

Top-bar wicker hives began to be replaced by modern frame hives during the third decade of the 20th century, but the vast majority of them were replaced in the 1950s and 1960s, especially in the latter decade. There were, however, voices insisting on the need to preserve the older hives. The most powerful of these was that of the agronomist and Head of the Department of Beekeeping at the Ministry of Agriculture, Panagiotis Georgantas, who, in a circular dated 1963, proposed not to replace them because, in his opinion, they offered significant quantities of honey and wax, while at the same time they were a source of replenishment of the damage to apiaries by pesticides, diseases, and natural disasters. Georgantas' encyclical did not ultimately have the effect that its author expected; on the contrary, it became the subject of fierce criticism from other beekeeping scientists. Thus, the top-bar wicker hives, which maintained millennia of beekeeping knowledge regarding the method of movable combs, being no longer able to respond to the demands of the times, were abandoned.

1.1.2 Skeps

1.1.2.1 *The migratory skep of northern Greece*

In some parts of northern mainland Greece and on some islands, there appeared a skep with an almost cylindrical shape and a more or less flat roof, which had a woven knob in its center (FIG. 29). This skep probably arose as a result of migratory beekeeping, which was employed by its users, for reasons related to loading and stacking. In sea transport, its shape helped in proper stacking, since the skeps were transported in rows and in a horizontal position on the deck of the vessels. Such placement would not be possible, for example, if the skep was conical. Similarly in land transport, its shape was helpful when loading on pack animals where, due to the existence of the woven knob on its roof, they were easier to secure, while in cases of transport in carts, trucks or even in train cars, good stacking and therefore safer transport was also possible[118].

FIG. 29. Migratory skep of northern Greece (Photo: G. Mavrofridis, 2016).

It cannot be ruled out that certain practices of beekeepers may have played a role in the formation of the shape of the skep in question or, conversely, these practices may have developed from its very shape. The way of multiplication of bee colonies, for example, by removing parts of honeycombs from a skep and fixing them with the help of wooden needles (i.e. pointed at one end of the stick that pierces the skep) into a new one, is aided by the shape of the skep. In a conical skep this work would be more or less problematic, since the fixation of the parts of the combs, usually seven in number, could not be carried out in the upper part of the hive. The attachment also of all the combs of the skep, unlike the conical ones, on the roof seems to have provided additional stability to the combs – necessary because of the disturbance of handling during transport – and facilitated some manipulations[119].

I therefore assume that the appearance of the skep we are considering is related to migratory beekeeping and the beekeeping knowledge gained during its exercise. Otherwise, the start of the use of this skep should be sought in the period of the beginning of migratory beekeeping in the region. But when did this migratory practice of beekeeping begin? There is not much information, but what there is, I believe, can lead us to some conclusions.

In the middle of the 10th century, in 942 or 943, a superintendent named Thomas wrote a report on the marginal differences that arose between the inhabitants of Ierissos and the monks of Mount Athos, in which an explicit order is found that the former should not create *melissourgeia* (apiaries) in the area of the Athonian monks[120]. In 943, the final separation of the lands between the inhabitants of Ierissos and Athonian monks took place, in the minutes of which no clause on apiaries was included[121]. It is not clear from the text of the supervisor Thomas whether the apiaries mentioned concerned permanent or periodic establishments that is a practice of migratory beekeeping by the beekeepers of Ierissos. If the latter was the case, then one cannot exclude that the hive used would have been the almost cylindrical skep of northern Greece. It should be noted that migratory beekeeping has been known in Greece since antiquity[122], while the area of Mount Athos was one of the main destinations of traditional beekeepers who used skeps, at least in the late 19th and 20th century when information is available[123].

The first definite reference to a migratory practice of beekeeping in northern Greece is found in a letter written by a French merchant named Rousset on August 7, 1716. In the letter, which is addressed to the consul of France in Kavala, Pierre Granier, a word is said about ships from Thassos that were anchored, in order to collect and transport the hives of the island as they did every year[124].

Indirect evidence of the existence of migratory beekeeping in Chalkidiki at the end of the 18th century is contained in a report of the French consul in Thessaloniki, Felix Beaujour. In this report, which the French diplomat composed on June 20th, 1797, it is stated that "Chalkidiki produces every year 30 to 40,000 okas[125] of wax. The island of Thassos does 25,000..."[126] (1 oka was 1.282 kg). Taking into account that migratory beekeeping was practiced in Thassos at the time, it seems most likely that the reported large wax production in Chalkidiki also came from migratory practice of beekeeping[127]. Besides, the existence of migratory beekeeping in the area is confirmed by another source, only a few decades later.

In 1830, Chalkidiki was visited by the English diplomat, David Urquhart, who in his travel impressions refers to the exploitation of honeydew in the pine forests of the area, and the transports carried out by the local beekeepers, moving their hives by sea to the shores of the Strymonic Gulf, as well as to the Singitic, Toronean and Thermaic Gulfs[128].

A little later, in 1845, in an Ottoman tax register concerning Liarigovi (nowadays Arnaia), 72 families of this settlement appear to be engaged in beekeeping having in their possession more than 2,500 hives. Each family owning between 20 and 100 hives. In the same document it is stated that 85 families of the village had mules, amounting to 103[129]. It can be reasonably assumed that a large number of these animals, as well as some of the other equines held in the settlement, were used for the transport of hives. As we know from the first half of the 20th century, the beekeepers of northern Chalkidiki kept animals, especially mules, for their needs. Often the only agricultural activity they engaged in involved the cultivation of foodstuffs for their animals[130].

In 1858, the German archaeologist and traveler, Alexander Conze arrived at the island of Thassos, and in his book mentions that, in June, the Thassian beekeepers transported their skeps to the opposite mainland coast, in Karaağaç (modern-day Porto Lagos) and after two months they brought them back to the island[131].

In 1861, the English traveler, William George Clark passed through Ierissos, and shortly afterwards published, under his initials (W.G.C.)[132], his travel impressions. Clark was accommodated in the house of Anagnostes Marinos, a village notable[133], who was absent at the time, because he had gone to see his hives in Thassos[134]. This information was noted by Clark in his diary, on September 13, a fact that makes clear that Marinos had taken his hives to Thassos for the exploitation of honeydew of the insect *Marchalina hellenica* in the pine forests of the island[135].

Indirect but clear evidence for the existence of developed migratory beekeeping in Chalkidiki, is the imposition in 1873 by the Ottoman administration of a tax on the transport of hives, according to which, one piaster should be paid per transported hive. Three years later, in 1876, a protest by the inhabitants of Liarigovi was published in the *Hermes* newspaper of Thessaloniki, where the dramatic situation in which beekeeping has fallen is deplored, due to excessively high tax and the arbitrariness of tax collectors[136]. For the record, this tax was eventually abolished, the exact date of the abolition being unknown.

From the above evidence, it follows that the migratory skep of northern Greece was in use at least from the beginning of the 18th century, and probably much earlier. Besides, this was the only type of traditional hive that has ever been recorded in the area of Thassos and Chalkidiki. However, the earliest description of this skep dates back to the late-19th century and comes from the English traveler Henry Fanshawe Tozer. He visited Thassos in 1879 and after eleven years he published a book, in which he describes the hives of the island as "cylindrical baskets of wickerwork, with a covering of earth, and a large stone on the top of each"[137]. From the cylindrical shape and the placement of stone on the top, it becomes clear that the skep used was the migratory one of northern Greece[138].

With this migratory skep the Thassian beekeepers moved their bee colonies to the opposite shores in the first half of the 20th century. The areas where they moved in spring – early summer were: Mount Athos for the exploitation of flowering (but also honeydew) of chestnut tree (*Castanea sativa*), as well as that of tree heath (*Erica arborea*); the plains of Xanthi and Chrysoupolis, through the ports of Porto Lagos and Keramoti respectively, for the exploitation of the flowering of Christ's Thorn (*Paliurus spina-christi*) and various herbaceous plants; in the area of Orfano for the same blooms and on the island of Samothrace, since the 1920s, for the exploitation of the flowering of the local oregano (*Origanum* spp.)[139]. Later, they also transported hives to the plain of Philippi[140]. Towards the middle of summer the hives were transported back to the island for the local flowering of various Lamiaceae plants and for the honeydew of the insect *Marchalina hellenica* in the pine forests of the island[141]. For wintering they moved most of their hives to leeward spots with plenty of autumn heath (*Erica manipuliflora*) around the Strymonic Gulf (MAP 3)[142].

The transportation was carried out on the island with pack animals to the beach and then with sailing vessels. In fact, some of the vessels of the island, such as those of Kakirachi (nowadays Kalirachi), were kept almost exclusively for the transport of hives[143]. Motor vessels were later used for maritime transport. On the opposite shores, when the area of hive placement was far from the beach, pack animals, usually mules, or carts were used. Later, mainly from the 1930s, trucks began to be used[144].

From the information of the Greek linguist, Georgios I. Kourmoulis (1907-1977), who visited Thassos in 1937 to collect linguistic material, and made a detailed ethnographic record[145] (still unpublished) of the island's beekeeping, we learn that the beekeepers of Thassos knew how to multiply their bee colonies in the spring without capturing the swarms. Firstly, they transferred a part of the bee colony with the queen to an empty skep and then from the original skep they cut half of each comb. These halved combs were placed with wooden needles in a third skep (FIG. 30) in which they shook the bees with the queen[146] from the empty skep. The skep with the queen was placed where the original hive was, while the one without the queen was transferred to some other area.

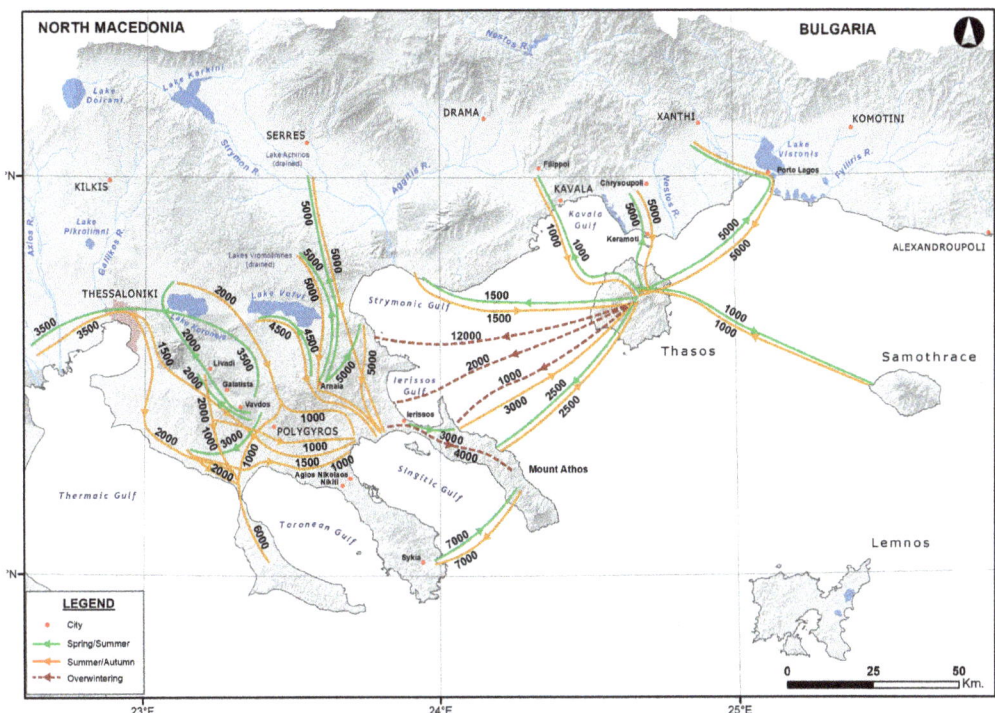

MAP 3. Map showing the migration of hives by the beekeepers of Thassos and Chalkidiki (except those of the Kassandra Peninsula and the villages of Nikiti and Agios Nikolaos) during the third and fourth decade of the 20th century (indicative number of hives per destination) (created by G. Tataris).

The traditional beekeepers of this island had the knowledge to lead the queenless hives to acquire a new queen. They cut a piece of comb with a queen cell or brood from another hive and placed it in the queenless one. Also, when they found that if a bee colony was weak, they changed its position with that of a strong bee colony to gain roughly the same population, or shook into the weak colony bees from a strong one. To prevent the bees killing each other, the Thassian beekeepers sprayed them with *raki* or anise-flavored alcoholic drink (which they poured with their mouths) or dusted them with corn flour[147]. With these practices the bees did not fight each other, and it was thus possible to introduce a new population to the weak bee colony. The Thassians were one of the few groups of traditional beekeepers in Greece who knew this method[148].

However, although the beekeepers of Thassos knew how to unite bee colonies, they did not use this knowledge solely to unite weak colonies to prevent their destruction. They also applied it during harvesting in order to achieve the maximum possible gain. They specifically gathered, through the practices just mentioned, the bees of two or three skeps-to-be-harvested in an empty skep, imprisoning the queens in special reed cages. Then, they cut off from the hives all the parts of the combs that had honey, while they adapted, with wooden needles, those that had brood to another skep, into which they finally introduced the bees and a queen. In this way, and by often carrying out three harvests a year, Thassian beekeepers received excellent yields: 13 to 20 okas (16.5 to 25.5 kg) or more honey from each skep – depending, of course, on the year; 0.8 okas (1 kg) of wax and 0.4 okas (0.5 kg) of *mudovina*, i.e. an alcoholic beverage made from the honey residues in the combs. But they only had a third of the original bee colonies left[149].

FIG. 30. Representation of placing parts of honeycombs in skep with the help of wooden needles (Photo: G. Mavrofridis, 2016).

So, having been left at the end of the year with only a third of his original bee colonies, the Thassian beekeeper usually transferred the weakest colonies to areas with plenty of autumn heath outside the island – around the Strymonic Gulf, as has been mentioned, – since the wintering on the island was far from ideal, mainly due to climatic conditions[150]. With the new beekeeping year, he was able to double his hives, in the way mentioned above, but he still lacked another third to make up for all the bee colonies of the previous year. This was covered by the purchase of bee colonies from Chalkidiki.

The supply of bee colonies from Chalkidiki seems to have been economically advantageous for the Thassian beekeepers because the price of products that could be produced from a bee colony – always in the special way they practiced beekeeping – was many times the purchase price of the colony[151]. It should be noted that bee colonies were bought with their skeps, and in this way the Thassians were freed from the whole process of making skeps or their separate purchase.

The sale of bee colonies by the beekeepers of Chalkidiki, especially those from the north part of the peninsula, in the area of Liarigovi, was also advantageous. Northern Chalkidiki presents a paradox from a beekeeping point of view. While the area is not particularly suitable for beekeeping, due to the poverty of beekeeping plants, its inhabitants, however, have managed to develop this practice to a high degree by taking advantage of flowering or honeydew from other regions and developing that type of beekeeping called reproductive, that is, beekeeping for the production of bee colonies for sale to other beekeepers[152]. The migratory skep of northern Greece itself was also considered ideal for this type of beekeeping. Its advantage was that its shape contributed to the concentration and preservation of heat in the winter months, which is why swarms within these skeps overwintered in the most favorable conditions and grew rapidly in the spring, to the highest degree allowed by their limited space[153].

With their knowledge, the beekeepers of northern Chalkidiki reached the point of increasing their bee colonies sixfold within a year, and under favorable conditions even increasing them eightfold (FIG. 31, 32), although the latter case was possible only in exceptional cases[154]. It seems that they relied a lot on the sales of hives to their Thassian colleagues, and they had a basically stable income.

For the purchase and sale of the bee colonies, special bee fairs were held twice a year, each lasting about two weeks, to which the beekeepers of Thassos came by sea. The first took place in the spring (April) in Ierissos, and the second in the summer (July) in Libsasda, nowadays Olympiada[155].

The earliest evidence of these bee fairs dates back to 1878[156]. The second testimony comes from 1887, where we learn of buying and selling of 6-7 thousand hives during the fair in Libsasda[157]. Also, in the *Pharos tis Makedonias* (Beacon of Macedonia) newspaper of Thessaloniki, dated July 30, 1888, it is mentioned that, in the village of Liarigovi alone, more than 1,500 Ottoman liras was gained from the sale of hives to the Thassians[158]. In the 1920s, Typaldos-Xydias talks about buying and selling of some ten thousand bee colonies in the two bee fairs of Chalkidiki[159].

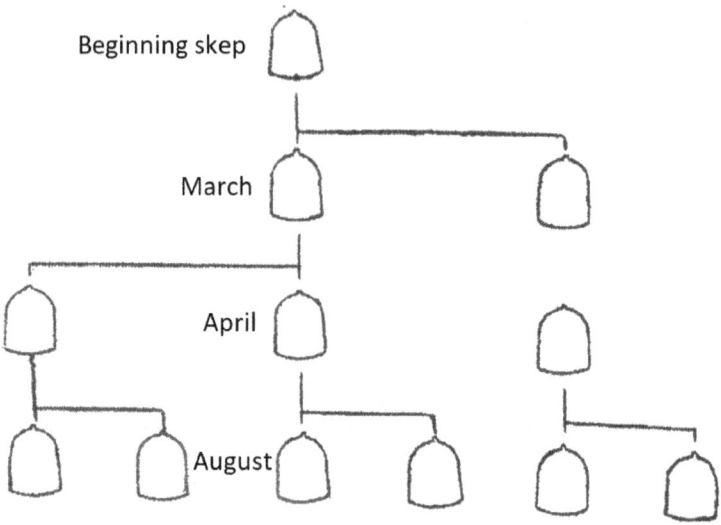

FIG. 31. Drawing showing the sixfold increase of hives within a year by the beekeepers of Chalkidiki (from Typaldos-Xydias 1927, 27).

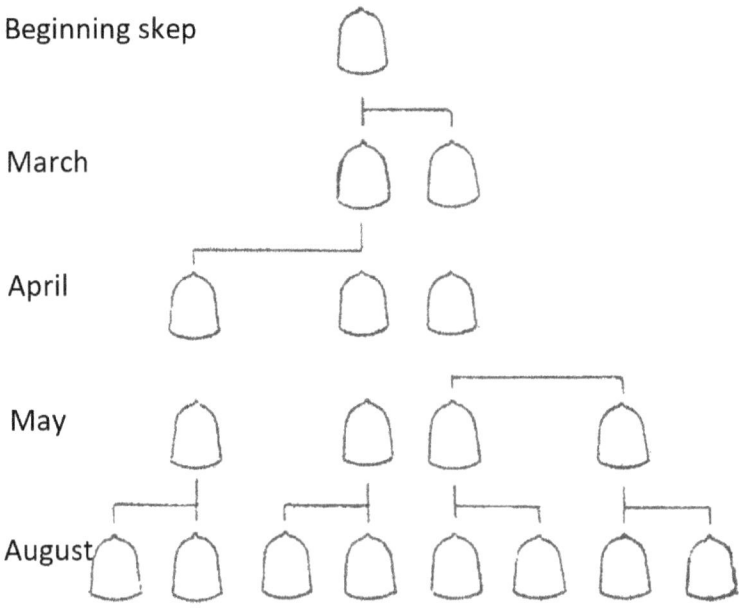

FIG. 32. Drawing showing the eightfold increase of hives within a year by the beekeepers of Chalkidiki (under favorable conditions) (from Typaldos-Xydias 1927, 27).

But when did the beekeepers of northern Chalkidiki start reproductive beekeeping? I believe that undoubtedly, both the beginning and the subsequent widespread development of their reproductive beekeeping is due to the need of their Thassian colleagues to procure bee colonies, as has been mentioned above. The benefit was mutual, and it seems that the cooperation of these two regions greatly benefited their beekeeping, to such an extent that they are the only areas of Greece where professional beekeepers, that is, beekeepers who had beekeeping as their only or main profession, made their appearance during the traditional beekeeping period[160].

In any case, during the 1870s, the reproductive beekeeping of the people of Chalkidiki and the purchase of hives by the Thassians through bee fairs was an established practice. It could therefore be reasonably assumed that this practice was an older one, although, in the absence of data, one cannot say how much older that was. However, the great development of beekeeping in Thassos at the end of the 18th century, which is inferred from the data cited by Beaujour, as well as the existence of a large number of beekeepers in Liarigovi in 1845, are quite strong indications that the reproductive beekeeping for sale to the Thassians in Chalkidiki was practiced from the middle of the 19th century, and probably even from the end of the 18th century[161].

The Chalkidiki beekeepers also practiced migratory beekeeping. During the first decades of the 20th century, they transported their skeps either by pack animals, by vessels, or by combined different forms of transport.

With pack animals, mules as a rule, hives were transported in the spring around the lakes of Achinos, Vromolimnes, Volvi and Langadas for the flowering of Christ's Thorn and various herbaceous plants; to the meadows of the lower course of the rivers Aliakmon, Loudias and Axios for the flowering of the various herbaceous plants; on Mount Athos, in the village of Livadi, and in the northeastern Chalkidiki for the chestnut forests; and to the southwestern coastal plain of Chalkidiki for the flowering of herbaceous plants, fruit trees and melons (MAP 3)[162].

Ten or twelve skeps were loaded on the mules, depending on the animal. Three skeps on each side of the animal, and four or six, in a horizontal position, on the back (FIG. 33), with their combs however in a perpendicular position to the ground. Transfers were almost always carried out during the night, and the beekeeper proceeded on foot (FIG. 34, 35). Mules usually belonged to the beekeeper himself[163]. Often, small groups of beekeepers traveled together. When the distance was too great to cover overnight, they would unload the skeps during the day and let the bees forage. When it was evening, they would reload them and move to the final destination. In this way they covered, in some cases, a distance of 100 and sometimes 150 kilometers in the course of three or four nights[164]. When they finally set up their apiary, they negotiated with the local farmers the supervision of the hives, and in return they offered them a quantity of honey[165].

FIG. 33. Representation of how skeps were loaded on a mule saddle by beekeeper, Asterios Giouvannakis (Photo: G. Mavrofridis, 2016).

FIG. 34. Mule loaded with skeps in Arnaia, Chalkidiki in 1926 (Photo: A. Typaldos-Xydias, from *Arnaia* 1996 9, 32).

FIG. 35. Christos Trikaliotis, a beekeeper from Arnaia, with skeps loaded on mules in the 1920s (Unknown photographer, from *Arnaia* 1996, 9, 32).

In spring, skeps were transported by sea from the Peninsula of Kassandra to ports in western Thermaic Gulf. From there the hives advanced by various means, usually pack animals or carts, to flowering herbaceous plants, as in the coastal plain north of Peneus River; in the meadows of the lower course of the rivers Aliakmon, Loudias, and Axios; in the plain of Axios; and around the Lake of Giannitsa. In addition, hives were also moved to the foot of Mount Ossa, for the exploitation of flowering in the forests of lime trees and chestnut trees (MAP 4). From Sykia of Sithonia the skeps were transported by sea to the chestnut forests of Mount Athos (see MAP 3)[166].

Sea transport, in case of calm and when the vessel was a sailboat, involved the risk that the bee colonies would die from asphyxia, due to the high temperature that developed in the closed hives. The placement of the skeps on the deck of the ships, as well as on the carts where they were used, was horizontal, with their honeycombs always in a vertical position to the ground. Care was also taken to keep the lowest possible temperature on the journey. Thus, the skeps were placed in rows, but their bases were not opposite each other so as not to increase the temperature of the bees (FIG. 36, 37)[167].

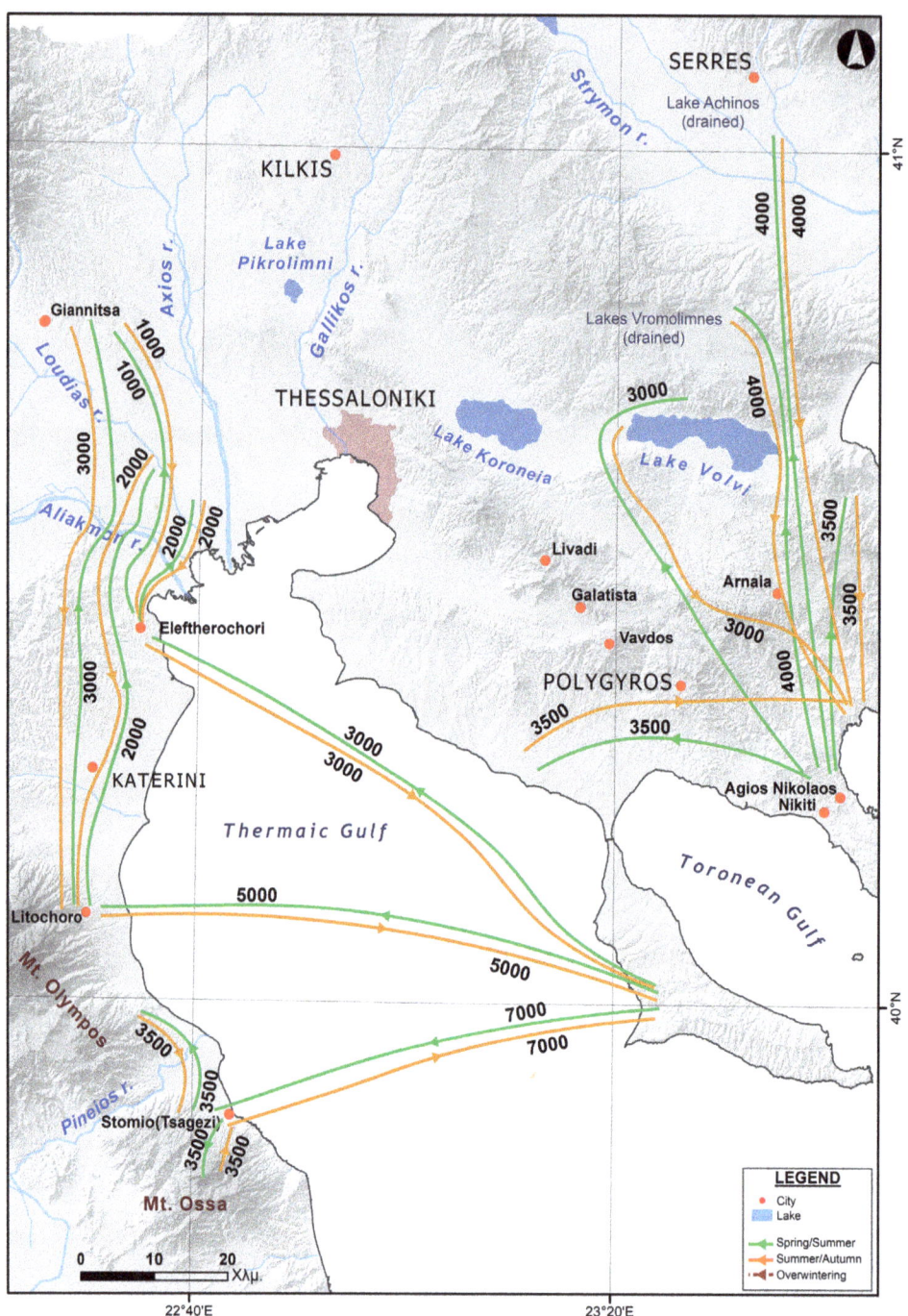

MAP 4. Map showing the migration of hives by the beekeepers of the Kassandra Peninsula and the villages of Nikiti and Agios Nikolaos in Chalkidiki in the third decade of the 20th century (indicative number of hives per destination) (created by G. Tataris).

FIG. 36. Representation of how the first rows of skeps were placed on the deck of the vessels (Photo: G. Mavrofridis, 2016).

FIG. 37. Representation of how skeps were stacked during their transport by vessels, carts, or train cars (Photo: G. Mavrofridis, 2016).

In August, the hives were transported back to Chalkidiki for honeydew in the local pine forests, those of Kassandra Peninsula and the northern part of Sithonia Peninsula[168]. The only exceptions were the beekeepers of Ierissos, Gomati and the surrounding area, who used to transport their hives in August to the island of Thassos. Apart from the easy access from their area, another reason was the expectation of greater honey production as was usually the case in the pine forests of this island[169], which consist of East Mediterranean pine (*Pinus brutia*) in contrast to those of Chalkidiki which are of Aleppo pine (*P. halepensis*).

In the areas close to the pine forests there was usually plenty of heath, so the bee colonies remained there to get stronger and overwinter[170]. Some beekeepers, however, especially those of Ierissos, transferred their skeps to Mount Athos, after returning from Thassos, as did many from Arnaia, Stanos and the surrounding villages[171].

The beekeepers of Chalkidiki usually carried out two harvests and under favorable conditions three in the year. The first harvest took place in the areas of spring-summer forages. The second took place in the pine forests as did the third. In the latter, if possible, only the strong hives were harvested, which, rarely exceeded half of the total number. Harvests were always done in the countryside. The beekeeper drove the bees of the colony-to-be-harvested, using smoke and blows on the outer walls of the skep, into an empty skep. Then he removed the combs and, with the help of wooden needles, adapted any parts bearing brood to a third skep into which he shook the bees that had been waiting in the empty skep[172].

The beekeepers of Chalkidiki knew how to multiply their hives, following a similar procedure to that of their Thassian colleagues[173].

On the skeps of Chalkidiki a layer of sprigs of various plants or grasses was usually placed for protection from the weather, often with a stone upon them[174]. A similar practice was used on Thassos[175]. In recent years, however, there prevailed in Chalkidiki the practice of placing on the skeps, as a cap, an empty, multi-layered, thick paper cement sack. This also applies to the few skep apiaries that exist today in Chalkidiki; in some cases a second sack (FIG. 38), of different material, is placed on top.

The skeps in Chalkidiki were of two sizes, which usually had differences in their construction materials. Those of Kassandra Peninsula, Mount Athos, and the seaside areas were larger, with a capacity of about 40 liters and indicative dimensions of 50 cm height, 30 cm diameter at the height of the roof and 35 cm at the height of the base. Their weight reached empty 5 okas (6.41 kg). In the north of Chalkidiki, smaller skeps were used, with a height of 40 cm, a diameter on the roof of 25 cm and at the base 28 cm. Their capacity was about 22 liters. In the former, rods of chaste tree (*Vitex agnus-castus*) were used for weft, while in the second, reed splints, which may not have been as durable but were much lighter, were chosen. The result of the smaller size and the use of reeds was that the skeps of north Chalkidiki weighed almost half of those of the coast, just 3 okas (3.85 kg) each[176].

FIG. 38. Apiary of skeps in Chalkidiki in 2016. For protection, they carry two types of sacks (Photo: G. Mavrofridis, 2016).

The reason for the variation in size and weight of the skeps had to do with the way (migratory) beekeeping was practiced in Chalkidiki. The beekeepers of northern Chalkidiki transported their hives almost exclusively with mules, and it was important that the skeps were as small as possible and did not weigh too much. On the contrary, those of Kassandra and the coasts carried out transport mainly by sea and sought long-term durability without caring about the larger size[177]. However, it seems that there was another reason for the use of the smaller type of skeps by beekeepers who practiced reproductive beekeeping. This was that the limited space encouraged the rapid growth of the bee colony[178], which was desirable. It also meant that when the skep was sold it looked full of bees and honeycombs (having, in fact, fewer bees and smaller honeycombs than the larger skep)[179].

As shown before, Thassians bought bee colonies mainly from the beekeepers of Liarigovi and the surrounding villages, who employed smaller skeps, and thus the smaller and lighter type was also used in Thassos (FIG. 39). Certainly, the island did not lack the largest skeps, but these constituted a very small part of the whole[180].

On the island of Skyros, the migratory skep of northern Greece was also used. However, it was constructed larger (FIG. 40, 41, 42), even by the large type of Chalkidiki, reaching a height of 75 cm with a diameter at the base of about 45 cm. The beekeepers of Skyros also practiced migratory beekeeping, but within the boundaries of their island.

Beekeeping With Woven Hives In Greece Through The Ages - Georgios Mavrofridis

FIG. 39. Skep on Thassos Island ("Thanassis Bikos" Photo Archive, Institute of Agricultural Sciences, Athens).

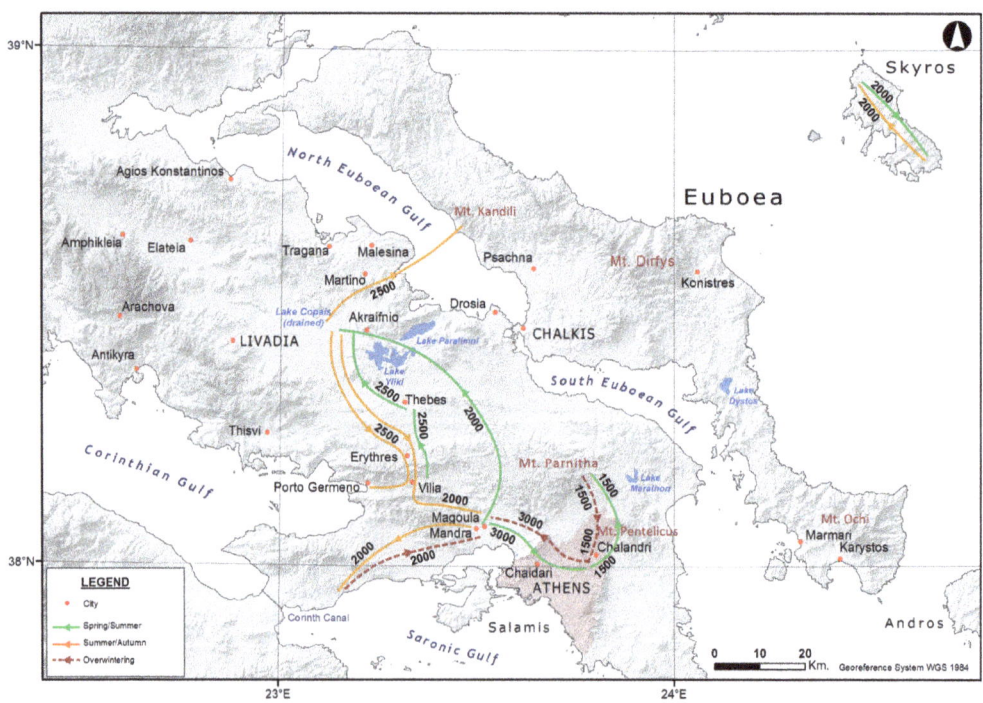

MAP 5. Map showing the migration of hives by the beekeepers of Skyros Island and their colleagues of Attica and Boeotia in the 1920s (indicative number of hives per destination) (created by G. Tataris).

Transport was carried out either by mules, on which up to seven skeps per animal were loaded, or by sea with *caiques* or fishing boats. In the spring the hives were transported to the southern part of the island for the flowering of sage (*Salvia fruticosa*) and thyme (*Thymbra capitata*)[181]. Later, the skeps moved to the pine forests on the north side, where there were also significant areas with autumn heath that allowed the bee colonies, after the pine honey harvest, to grow stronger, in order to overwinter (MAP 5)[182].

FIG. 40. Skep from Skyros Island (Photo: G. Mavrofridis, 2019).

FIG. 41. Beekeepers of Skyros in their apiary, in 1960 ("Thanassis Bikos" Photo Archive, Institute of Agricultural Sciences, Athens).

FIG. 42. Traditional beekeeper of Skyros in his warehouse, in 1960 ("Thanassis Bikos" Photo Archive, Institute of Agricultural Sciences, Athens).

The beekeepers of Skyros used essentially the same methods and techniques as their colleagues of Thassos and Chalkidiki, with the exception of the union of bee colonies, which was applied only by the Thassians[183].

Here, the skeps were also covered though in a different way. On the roof of each skep they placed a thick layer of pine branches which were tied together. This layer supported a large oblong stone slab, which protruded noticeably from the skep. The skeps were placed in order, one next to the other, so that the oblong plates stepped onto each other thus supporting themselves, and a set was created that included all the skeps of the apiary[184].

The beekeepers of Skyros, like those of Thassos, also used queen cages made of reed (FIG. 43), while for the creation of smoke, they used touchwood as fuel, just like their colleagues from Thassos and Chalkidiki[185].

FIG. 43. Tools of traditional beekeepers of Skyros (Photo: T. Bikos, from Bikos 2008c).

The production in all three areas examined here, Thassos, Chalkidiki, and Skyros, was quite large and while the gathering of honey was relatively easy[186], for wax, which in the past was considered a particularly valuable material, special wooden structures were used, called *keromyloi* or "wax mills" (wax presses)[187]. The earliest evidence of the use of "wax mills" we possess dates back to 1849 and comes from Thassos. That year a contract of sale was drawn up for a house with a "wax mill"[188]. About twenty years later, the Monastery of Vatopedi on Mount Athos, according to the monastery's archive, built a "wax mill" in the village of Voulgaro (nowadays Rachoni) on Thassos, to which the local beekeepers went to receive, for a fee, their wax[189]. In Chalkidiki the owners of "wax mills" in the 1920s charged one drachma per oka (1.28 kg) of extracted wax[190]. The only wooden "wax mill" of these areas that has survived to this day is located in Skyros and is in private hands[191].

Except from the three main areas of northern Greece, the migratory skep was also found elsewhere in the mainland as well as on the islands, mainly where the beekeepers of Thassos and Chalkidiki transported their skeps. In these areas, local beekeepers, perhaps under the influence of their colleagues from Thassos and Chalkidiki, appear to have been using – since at least the middle of the 20th century – the same type of skep, while exercising, in most cases, static beekeeping. Such areas were the west coast of the Thermaic Gulf and some areas of the prefecture of Thessaloniki[192]. As far as the islands are concerned, the skep in question has been recorded on Skopelos[193], Skiathos[194] and Agios Efstratios[195].

From the second decade of the 20th century, the beekeepers of Chalkidiki who were engaged in reproductive beekeeping began to sell hives to beekeepers from other regions, not only to those of Thassos. During this period there appeared in the bee fairs of Chalkidiki beekeepers from northern Euboea and some villages of central mainland Greece buying hives[196]. These beekeepers used other types of traditional hives, but by buying bee colonies along with their skeps from Chalkidiki they transferred the use of migratory skep of northern Greece to their areas.

After the middle of the last century, the beekeepers of Chalkidiki, with their knowledge, managed, in the spring, to supply their colleagues throughout Greece with bee colonies, including the southernmost regions, even though the former overwintered their hives in the noticeably worse climatic conditions of their homeland. Thus, colonies that wintered in skeps in Chalkidiki, having subsequently been transferred into modern movable-frame hives, were sold all over the country. In some cases, however, the bee colonies remained in their skeps, and some beekeepers used them in turn, alongside their frame-hives, primarily for early spring colonies production. As a result, this skep appeared in use in areas where open-at-the-bottom traditional hives were completely unknown.

Typical examples are those of the eastern Peloponnese[197] and central-western Crete[198], where endemic hives were open-at-the-top or open-at-the-top-and-bottom with movable combs. The skep of Chalkidiki was used by some beekeepers during the 1970s and 1980s, not for traditional beekeeping in the area, but as a tool in modern beekeeping practice to produce early colonies in the spring. They are still to be found in Chalkidiki (FIG. 44), although there are now very few beekeepers who use the local skep for this precise purpose.

FIG. 44. Beekeeper, Christodoulos Giouvannakis from Gomati, Chalkidiki, inspects one of his skeps (Photo: G. Mavrofridis, 2016).

1.1.2.2 The migratory skep of Attica and Boeotia

In Attica and Boeotia, a skep appeared, similar in shape to the migratory one of northern Greece, with which migratory beekeeping was also practiced. This skep was, however, noticeably larger[199] (FIG. 45). In Attica, it was used by the beekeepers of the villages of Mandra and Magoula in Thriassio Plain, and by those of Vilia and Kriekouki (nowadays Erythrae). In Boeotia it was used by the beekeepers of some villages close to Kriekouki, such as Kokla (nowadays Plataea), Kapareli and Karanta (nowadays Ellopia)[200]. In some rare cases, static beekeeping was also practiced with this skep[201].

FIG. 45. Beekeeper with migratory skeps of Attica at the beginning of the 20th century (Unknown photographer, from Toufexis 1904, 85).

The large size of this skep was not a problem for the local beekeepers, because the existence of roads allowed them to use two-wheeled carriages for transport. These carriages were leased by the beekeepers along with their coachmen, who undertook the transport of the hives. In some cases, the transportation lasted up to four nights, while each carriage could carry 25 skeps. In some cases, mainly related to Kriekouki and the surrounding villages, pack animals were also used, which were kept by beekeepers for other agricultural work. During the 1920s, trucks also began to be used. On each of them, 50 skeps could be transported in one night, at a distance that required a travel of three nights by a couple of two-wheeled carriages[202].

The beekeepers of Mandra and Magoula, usually starting in June, moved their skeps to Skaramangas and the Athenian Plain (Dafni, Tourkovounia, and Chalandri) for the early flowering of thyme (*Thymbra capitata*). After about a month, they moved them to the foothills of Mount Parnitha and Mount Penteli, where, due to the altitude difference, there was late flowering thyme. Later, the bees exploited there the pine tree honeydew and the blooms of the autumn heath (*Erica manipuliflora*) and the strawberry tree (*Arbutus unedo*) (MAP 5)[203]. Regarding the size of migratory beekeeping within the Athenian Plain, an important report of 1907 tells of the placement of 3,000 skeps in only one of the destinations of the beekeepers of Mandra and Magoula, namely, that of Chalandri[204]. In the 1920s, however, the same number (3,000) applies to the total number of skeps transported to the Athenian Plain[205]. This is probably due to the wars in which Greece was involved between 1912 and 1922[206].

In cases of adverse conditions in the Athenian Plain, the beekeepers of Mandra and Magoula transferred their skeps to the area of the then partly-drained Lake Copais for the flowering of cotton and various local wild plants[207]. In some cases they moved skeps in autumn, into the pine forests east of the Corinth Canal[208].

The harvesting of the skeps was done on the spot, before each transfer, with the help of an empty skep into which the bees were driven, so that the beekeeper could harvest without their presence[209].

The beekeepers of the villages of Kriekouki and Vilia, as well as those of neighboring villages of Boeotia, transferred their skeps in July to the outskirts of Thebes for the flowering of thyme, and in August to the varied flowering of Copais. Then they moved them to the pine trees of the Corinthian Gulf, near the villages of Psatha and Germanos (nowadays Porto Germeno), for honeydew and overwintering. However, in the third decade of the 20th century, due to the fires in the aforementioned pine forests, several beekeepers of northwestern Attica and the neighboring region of Boeotia transported their skeps beyond Copais, with the help of small sailing vessels, to the foothills of Mount Kandili on the island of Euboea. This sea transport lasted no more than three to four hours, and the skeps that were transported numbered about 2,500. In the area of Kandili, the bees took advantage of the pine tree honeydew, and the blooms of the autumn heath and strawberry tree, and overwintered. In the spring of the following year, in March, beekeepers moved the skeps back to their site to start the annual cycle some three months later (MAP 5)[210].

A question that can be raised here is whether there were professional beekeepers in Attica and Boeotia, that is, beekeepers having beekeeping as their sole or main profession, as was the case in Thassos and Chalkidiki. From the available data, it seems that despite the relatively large production of honey (and therefore also of wax), there were no professionals among the beekeepers of Kriekouki and the surrounding villages[211] and the same was true of the beekeepers of Thriassio Plain[212]. The beekeepers of these regions were basically farmers and beekeeping was for them, as for almost everywhere else in Greece, a sideline profession.

1.1.2.3 *Conical and bell-shaped skeps*

Conical and bell-shaped skeps were used by the beekeepers of several regions. They were found in Pontus, western Asia Minor, Thrace, Macedonia, Thessaly, Central Greece, and some islands of the Aegean and Ionian Seas. They were also widespread in the Balkans[213].

FIG. 46. Old beekeeper, Yiannis Nikolaidis, with his conical skep in Koukos, Pieria (Photo: G. Mavrofridis, 2008).

In Pontus (nowadays Turkey), conical skeps seem to have been used in some areas, alongside the main type of Pontic hive, which was horizontal, made of tree trunk. Their use has been recorded in Larachani of Matsouka (in the Trabzon region), where, due to the humid climate, they were not placed on the ground or on a stone base but on a specially constructed wooden bench[214]. Conical skeps were also used in some cases by beekeepers of Pontic origin in their new homes in Greek Macedonia, such as in Koukos, Pieria[215] (FIG. 46), and Angistro, Serres[216].

In western Asia Minor, among other types of hives, skeps were also used. According to Eva Crane, the use of these skeps in western Asia Minor is due to Balkan influence. In 1985 Crane encountered in the Bursa region conical skeps, which were used by a Turkish beekeeper, whose distant origin was from present-day Bulgaria. In a photo she cites, the skeps are characterized as of "Bulgarian-type"[217]. Elsewhere however, Crane herself refers to another beekeeper using similar skeps, though she does not give evidence of his origin[218]. We know that in the 1930s conical skeps were widespread in northwestern Asia Minor (especially in the areas around Adapazar and Bolu)[219]. Besides, Brother Adam alias Karl Kehrle (1898-1996), encountered these skeps on his travels, both to the south of western Asia Minor, near Isparta[220], and to the Anatolian plateau, in an area some 30 km northwest of Ankara[221]. There is also a report that Greek refugees from Asia Minor brought the use of conical skeps of their homeland to the place of their new settlement, in the area of Kastoria[222], while it seems that they were also used by Greek beekeepers in Mechaniona (nowadays Çakilköy), in northwestern Asia Minor, Turkey[223]. Besides, the same skeps, as was already mentioned, were used in Pontus.

From the above it becomes clear that, apart from the isolated case recorded by Crane, there is no connection between the conical skeps used in Asia Minor and Bulgaria. Besides, even if we accept that the skeps of Asia Minor bear a Balkan influence, as Bodenheimer believes[224], the question is, why this must have been from Bulgaria and not from any other region of the Balkans, especially the neighboring Eastern (Turkish) Thrace[225], where conical skeps were predominantly used by both Greek (as we will see below) and Turkish beekeepers?

In Thrace, skeps seem to have been the only type of traditional hive, at least in the 20th century. In Eastern Thrace, they were conical or bell-shaped. In the village of Kryonero (nowadays Soğucak) they were made conical[226], but in other areas, such as the outskirts of Constantinople (nowadays Istanbul), the skeps seem to have had a vaulted roof that gave them the shape of a bell. Greek refugees from the region manufactured and used bell-shaped skeps in Florina, continuing the tradition of their place of origin (FIG. 47)[227]. Interestingly, in the area around Mount Strandzha, in Northern Thrace (Bulgaria), near the border with Turkey, skeps had also the shape of a bell[228]. In the rest of Northern Thrace, conical skeps were almost exclusively in use[229], and probably this type of hive was used in the past by the Greeks of the region. Greek refugees from

Bulgarian Thrace, who settled in 1928 in villages of Western (Greek) Thrace, used this type of hive[230].

FIG. 47. Bell-shaped skep from Florina Prefecture (Photo I. Anagnostopoulos, 1990s).

In Greek Thrace skeps were made conical (FIG. 48) or bell-shaped[231], except in the area around Didymoteicho, where they had a peculiar shape – that of truncated cone (frustum) that nevertheless ended in a pointed peak (FIG. 49)[232]. The Turkish-speaking Muslims of the region and the Pomaks (Slavic-speaking Muslims) also practiced beekeeping with conical-shaped skeps[233]. All the skeps of Thrace were anointed with a mixture of dung (which should come from a cow that had given birth!) and ashes. Some people added mud to the above mixture. In winter, the base of the skep was covered with the same mixture, but a hole was left for the bees to pass through, which should look to the East[234].

FIG. 48. Conical skep from the area of Orestias, Evros Prefecture (Photo: G. Mavrofridis, 2007).

FIG. 49. Skeps of peculiar shape from the area of Didymoteicho, Evros Prefecture (Photo: Folklore Museum of Didymoteiho, 2019).

For protection from the natural elements, the skeps in the whole of Thrace were covered with a straw hackle, while in many cases they were placed inside a fence made of braided branches[235] or a stone-built enclosure[236]. There is an interesting report, dating back to the 1620s, about the use of fences for hives in the area. It concerns Eastern (Turkish) Thrace and is a response to a letter sent on 16 December 1659 by John Winthrop the Younger (1606-1676), an early governor of the Connecticut Colony, to Samuel Hartlib (c.1600-1662), the great agricultural reformer of his age. There, reference is made to "a yard hedged in", which the author of the letter saw in the late 1620s not far from Constantinople (Istanbul). Within this fenced area, there stood a large number of hives stepped on stone slabs or boards[237]. Although Winthrop does not describe the type of hives, it seems most likely that skeps were the hives placed within the fence, by taking into account that they were the traditional hives of the area. The fact that the hives to which he refers stepped on stone slabs or boards strengthens this view, because the same materials were also used in Thrace to protect the skeps from moisture[238]. The conical or bell-shaped skeps would also be known to Winthrop from England, where they were the only type of hive in common use[239]. If he was referring to a type of hive unknown to him and the recipient of the letter, he would have made a more detailed description.

Beekeeping with conical or bell-shaped skeps in Thrace, as in all other areas of their use in Greece, required the hunting and capture of swarms in the spring, since the beekeepers who cultivated them did not have the knowledge to multiply their bee colonies by artificial swarming, as did their colleagues with the flat-roof migratory skeps, and others with the movable-comb wicker hives.

In Vrysika of Didymoteicho, in Western (Greek) Thrace, when a bee colony was swarming, they threw water or soil into the swarm, believing that the swarm might think it was raining and sit on a nearby branch. To collect it, they rubbed the inside of an empty skep with lemon balm (*Melissa officinalis*) holding it with the left hand, while with the right they held a bunch of the same plant. When they wanted to prevent swarming, they cut the queen cells[240]. They did the same in the nearby village of Koufovouno[241].

FIG. 50. Conical skep from Florina Prefecture (Photo: G. Mavrofridis, 2007).

If we exclude the areas in northern Greece where the migratory skep was used, in the rest of Macedonia, the skeps were in their vast majority conical. The use of these skeps has been recorded in many prefectures of the area, such as Drama[242], Serres[243], Pieria[244], Grevena[245], Kastoria[246] and Florina[247] (FIG. 50). However, it must have also been used in the other prefectures of Greek Macedonia, with the obvious exception of Chalkidiki.

In the north of the prefecture of Serres, beekeeping with conical skeps (FIG. 51, 52) seems to have been highly developed. There were beekeepers, as for example in Achladochori, with a large number of bee colonies in skeps, which in some cases exceeded two hundred. These skeps were often transported, but these journeys were exclusively local in nature within areas around the village[248].

FIG. 51. Conical skep in Sidirokastro, Serres Prefecture (Photo: G. Mavrofridis, 2008).

FIG. 52. Apiary of skeps in Sidirokastro, Serres Prefecture (Photo: G. Mavrofridis, 2008).

The skeps in the prefecture of Serres were usually placed on stone slabs or stones for protection from moisture, while in winter the opening around the base of the skep was closed with mud, except for a small point that served as an entrance. For protection from the weather, a straw hackle was often placed on top of the skep[249].

In the prefecture of Kastoria, in Western (Greek) Macedonia, the skeps were slightly larger than those of Serres. Peculiar to the area was the use of one and sometimes two crosses made of rods that pierced the skep and served to better fix the honeycombs. In addition to free placement in the bee enclosures, the skeps were also placed in bee boles, which existed in the courtyard walls of the houses[250].

In the prefecture of Kastoria, as well as in the neighboring prefecture of Florina, the skeps were covered with a straw hackle. For its construction the straw was soaked to be malleable, knotted, and then turned over to find the knot from the inside (at the bottom). On top of this hackle, they then placed an old military helmet! In Florina, beekeeping, which was usually practiced in the context of the domestic economy, was exercised mainly by women[251].

Conical skeps were also found in Thessaly, in the prefecture of Larissa. Their use has been recorded in the area of Elassona[252], and in the villages of the plain south and east of the city of Larissa[253].

Further south, in Central Greece, conical or bell-shaped skeps were used in many areas. In Attica and Boeotia, these skeps were covered with a straw hackle, the straw of which, however, was cut above the point used for entry by the bees (FIG. 53) to facilitate, as they believed, the orientation of the bees[254].

FIG. 53. Conical skep in Boeotia with a peculiarly "clipped" straw hackle
(Photo: P. Papadopoulo, 1937).

In 1842, Joseph Philibert Girault de Prangey (1804-1892) from France, one of the pioneers of the art of photography, visited Attica. By using the technique of daguerreotype, a forerunner of photography, he immortalized the most important ancient and Byzantine monuments of the city, among which was the Monastery of Dafni. One of the three daguerreotypes of this monastery shows part of the east side of the destroyed church[255]. The interesting thing is that in front of the wall of the catholicon, a beekeeper had placed three skeps, which were also immortalized in the daguerreotype.

FIG. 54. Skep in a bee bole on Andros Island (Photo: G. Mavrofridis, 2021).

Conical skeps were also used by some beekeepers in the mountains of Central Greece. However, there the skeps seem to have constituted only a small percentage, not more than one tenth, of the entire traditional hives[256]. Hives multiplied exclusively by capturing swarms in the spring[257].

Conical or bell-shaped skeps were also used on some islands. One of them is Andros. Here they employed both kinds of skeps[258], but it seems that the bell-shaped ones were preferred above the others[259]. The skeps were placed into bee boles created in the dry-stone walls of the terraces (FIG. 54) or in those separating the various properties. In the area around Korthi, in the southern part of the island, a stone slate was often placed at the opening of the bee bole and in front of each skep, for protection (FIG. 55)[260].

FIG. 55. Skep on Andros in its bee bole with a stone slab in the front of the bole for additional protection from the natural elements (Photo: G. Mavrofridis, 2021).

Externally, and in many cases internally, the skeps were anointed with a mixture that included cow's dung, and often other materials, such as ash, red soil and lime, depending on the village and the practices followed by each beekeeper[261].

Several of the beekeepers of Andros adapted into the skep, just before the roof, a small cross of twigs or a thick wire so that the bees could start building their combs from there[262].

Hornets (*Vespa orientalis*) were a major problem for beekeepers on the islands of the southern Aegean, and it was not uncommon for entire apiaries to be destroyed by them. To help bee colonies in dealing with them, the beekeepers of Andros took various measures. One of them was to reduce the size of the entry of their skeps so that these large insects could not enter. This was achieved by placing a special stone "comb" at the entrance[263], through the openings of which the hornets could not pass, or by placing a bunch of sprigs of small bushes[264]. Other measures taken were the use of hornet traps, the killing of queen hornets before they could create a colony, the destruction of the hornet nest, and the killing of hornets with brooms and branches in the vicinity of apiaries. Finally, there were some who went to the apiaries for work; they killed the hornets and received honey as remuneration from the beekeepers[265].

Some of the users of skeps on Andros, probably not satisfied with the results of static beekeeping, practiced, with their skeps, migratory beekeeping, but within the boundaries of their island. Andros was the only area in Greece where migratory beekeeping was practiced with conical and bell-shaped skeps. The transportation was carried out at night either by the beekeeper himself, who carried two skeps tied together with rope – one on his back and one on the chest, or with donkeys[266].

The honey yield of the island's skeps ranged from two to ten kilograms. Harvests in the past were carried out up to three times a year: the first in May/June for honey derived from wildflowers and citrus fruits; the second in August for thyme (*Thymbra capitata*); and then in autumn for heath (*Erica manipuliflora*) honey[267].

Women on Andros were not allowed to go to the apiaries or practice beekeeping themselves. The beekeeper, on the other hand, should not have slept with a woman the day before his visit to the apiary[268]. The beekeeper should also not have slept, in the night before visiting the hives, in the same bed as his wife when she had her menstrual period (she had to sleep in another room of the house!), while it was believed that bees would sting the beekeeper even more if he visited them after sexual contact with a foreign woman[269]. Similar superstitions were in force in antiquity. They are also found on other Cycladic islands, as well as on Rhodes[270]. There were, however, women who, either out of necessity or otherwise, dealt with hives. One of them, coming from the island of Tenedos (nowadays Bozcaada, Turkey) but married on Andros, remained a young widow and practiced beekeeping exclusively with skeps, which she knew from her island[271].

The above indirect evidence is the only one known of the traditional hives used on Tenedos. It seems that similar skeps were also in use on Tenedos. On the rest of the Aegean islands the conical or bell-shaped skeps were unknown[272]. This type of traditional hive, however, was found in the Ionian Sea, on Corfu, and specifically in the village of

Perivoli[273], in the southern part of the island. However, the only thing we know about them is that they were used until about 60 years ago.

So far, we have examined the areas where conical or bell-shaped skeps were found, as well as some of the methods of beekeeping associated with them, but we did not refer to the way in which the harvest was carried out, which was one of the main differences in the various regions. There were three ways of harvesting. The first involved killing the bees and taking all the honeycombs of the skep. It was applied by the majority of beekeepers in Thrace and Macedonia, except for the prefecture of Serres, as well as by beekeepers in some mountainous regions of Central Greece. During this practice, some skeps were left unharvested to become the "mothers" who would give the swarms of the new year[274]. In the rest, the bees were killed, and the empty skeps were placed for protection in a usually covered area until the next spring, when the beekeeper would introduce the swarms he captured, and the new beekeeping year would start. The killing of bees was carried out either by drowning the bees in water[275], by trampling them (after sprinkling them with flour water so that they could not fly)[276], or finally by fumes from sulfur[277].

In the second way of harvesting, the beekeeper would turn the skep sideways or upside down, smoke the bee colony, and cut off the most of combs containing honey. This method was used by the beekeepers of Thessaly[278], Andros[279], Serres[280], and an area of the prefecture of Kastoria[281].

The third and most rational way of harvesting was found in the beekeepers of Attica and Boeotia[282], as well as in some of their colleagues in Serres (Angistro)[283] and Evros (Gemisti)[284]. This was almost the same method applied by the users of the migratory skep of northern Greece, as well as those of the migratory skep of Attica and Boeotia. According to this method the beekeeper drove, with the use of smoke and blows on the outer walls of the skep, the bees to an empty skep, which he had placed at an angle to the skep to be harvested, while a white cloth played the role of the bridge between the two skeps. When the bees passed into the empty skep, he cut off the combs with honey, leaving some of them the overwintering of the bee colony. When it was over, he repositioned the bees in their original skep.

Nowadays, conical and bell-shaped skeps are rarely used, but a small number of beekeepers, especially in Thrace, employ them alongside their frame hives (FIG. 56).

FIG. 56. Conical skep from Thrace for sale at the 2018 Honey Festival in Piraeus (Photo: G. Mavrofridis, 2018).

1.2 Horizontal Wicker Hives

The use of horizontal wicker hives was not particularly widespread among Greek beekeepers, but they were used in some areas where beekeeping was practiced with horizontal hives, such as the islands of the Aegean, as well as by refugees from Asia Minor who brought their beekeeping tradition to their new homes (in Greece). They were also known in Cyprus, where they were not widely accepted.

The Aegean islands on which the use of horizontal wicker hives has been recorded are Amorgos, Ikaria, Chios, Lesvos, Symi and Anticythera. For Amorgos there are reports that in the past, from around the end of the 19th century to the beginning of the 20th century, open-at-both-ends cylindrical wicker hives smeared with dung were used on the island[285]. On Symi, alongside the characteristic horizontal board hive of the island, there were also, in smaller numbers, open-at-both-ends hives of cylindrical shape, made of wicker, with a length of up to one meter[286].

On Anticythera, the use of horizontal open-at-both-ends wicker hives has also been recorded, alongside top-bar hives and wall hives[287]. The beekeepers of Lesvos[288] and Chios[289], in addition to various other types of traditional hives, also employed horizontal open-at-both-ends wicker hives. In all these cases, no wicker hives have survived, and no photograph of them is known to me. There is only one rough sketch of the horizontal wicker hives of Symi by Thanasis Bikos[290], based on the testimonies of his informants. Things are a little better as far as Ikaria is concerned. At least one horizontal wicker hive has survived and is exhibited in the museum collection of the local beekeeping co-op (FIG. 57)[291]. This wicker hive is not exactly cylindrical, as its mouth has a slightly larger diameter[292]. There, a stone slab was placed as a lid, which at one point of its circumference had a curved opening for the entrance of the bees.

FIG. 57. Horizontal wicker hive from Ikaria Island (Photo: Thanassis Bikos, 2000s).

On Ikaria, the traditional clay hives were open-at-one-end and were placed for protection within a peculiar construction of stone slabs that enclosed each hive[293]. In the case of this island, it seems that horizontal wicker hives were open-at-one-end just like clay hives. So the harvest, like all the management manipulations of the beekeeper, were carried out from the only mouth, which was also the bees' entrance. This was of course a disadvantage, because the honeycombs, from the middle to the back of the hives, were never harvested and therefore not renewed, with all that this meant for the longevity of the bee colony. On the other islands, where horizontal wicker hives were open-at-both-ends, the harvest was carried out alternately, from the front and the rear opening respectively, and the honeycombs were constantly renewed.

In Cyprus, where all types of traditional hives were horizontal, there were also wicker hives. The oldest evidence of their use on the island dates back to the 1870s. At the end of this decade, only a year after Cyprus passed into the hands of the British Empire (1878), the Italian beekeeper and traveler, Giuseppe Fiorini visited the island. The purpose of his visit was to procure hives of the local bee subspecies (*Apis mellifera cypria*) and to create a unit for sending bees (queens) to Europe. What interests us here is that one of the eight bee colonies that Fiorini bought was in a horizontal wicker hive coated with clay. According to his words, the inside of the hive "was well polished and painted with a kind of white varnish. This paint helps the taking out of the comb, for by putting a knife between the combs and the hive the varnish is broken loose and the comb may be taken out whole"[294].

Use of horizontal wicker hive is also attested in Agios Andronikos of Karpasia, in the northern part of Cyprus. These wicker hives were open-at-both-end, cylindrical, and smeared, externally and internally, with a mixture of mud and chopped straw[295].

In Messokampos, in the prefecture of Florina (northwestern Greece), the use of horizontal wicker hives by refugees who settled in the area has been recorded. These wicker hives were used until the 1970s, while two have survived (FIG. 58, 59) and one is exhibited in the permanent beekeeping collection of the Folklore Museum of Florina. The refugees' place of origin was Ak Dag Maten (modern-day Akdağmadeni), east of Ankara. The Greeks who settled in Ak Dag Maten came gradually from Argyroupolis (nowadays Gümüşhane) of Pontus, between 1832 and 1880, and spoke the Pontic dialect[296]. Similar wicker hives were characteristic of the Highlands of Anatolia[297], and it was probably there that they began to be used by the Pontians of Ak Dag Maten, since their use was unknown to Pontus.

FIG. 58. Front part of the horizontal wicker hives from Mesokambos, Florina Prefecture
(Photo: I. Anagnostopoulos, 1990s).

FIG. 59. Backside of horizontal skeps from Mesokambos, Florina
(Photo: I. Anagnostopoulos, 1990s).

The wicker hives of Ak Dag Maten / Messokampos of Florina were open-at-one-end, about 85 cm long, and had the shape of a truncated cone. Their front, closed end, was 16 cm in diameter, and had a small hole for the entrance of bees. At the rear end, about 35 cm in diameter, a plank lid was placed, for the correct placement of which there was, as an indicator, a white line of lime which was aligned with the corresponding line that existed on the body of the wicker hive, near the lid. The wicker hives were anointed both externally and internally with a mixture of dung and ashes. Their installation was not done on the ground but on a wooden bench, 30-40 cm high, so that they did not rot quickly and to protect them from the winter cold. Above the wicker hives, they placed a sheet of metal upon which they placed bales of straw[298].

For the capture of swarms in the spring, a special small basket with a handle was used, which was sprinkled internally with warm water to increase the smell of the beeswax that was placed on its walls. This special basket was often hung on a tree, and thus attracted a swarm naturally. When beekeepers saw a bee colony swarming, they would pat their palms or have children singing in an attempt to guide the swarm to an accessible point for its capture[299]. The captured swarm was transported within the special basket and transfused into a horizontal wicker hive.

The hives were offered food after winter, during the first spring rains. Of interest was, however, the type of food that was given – a well-cooked chicken sprinkled with honey, as well as bread crumbs mixed with honey! A similar offering with a chicken placed in horizontal wicker hives took place during the winter, according to Eva Crane's oral testimony to Ioannis Anagnostopoulos[300]. The same practice has been recorded in northern Iraq. In both cases, certainly the chicken, and also the bread, was not eaten by the bees but by other organisms, but the offering seems to have been intended to declare the respect and gratitude of the beekeeper for what the bee offered him[301]. Similar reports of feeding bee colonies with poultry meat can be found in the works of Columella (*De re rustica* IX.14.15) and Pliny the Elder (*Naturalis historia* XXI.48) who lived in the 1st century AD.

However, chicken being offered to bees, particularly to the Asian species of *Apis cerana*, has also been recorded in China. It is mentioned in the early-16th century AD by the Chinese writers, Song Xu and Gui E[302]. David Pattinson, who studied, among other things, the work of these two Chinese writers, in his attempt to explain why chicken was offered as food, speculates that the widespread belief in China that chicken is particularly healthy for humans and this benefit was also transferred to bees, at least in some areas[303]. The offering of chicken meat to bee colonies is extremely interesting because, as is well-known, they do not consume food of animal origin. The fact that this practice was found in Roman antiquity, in 16th-century China, in modern Greece and the Near East is striking and deserves, in my opinion, further investigation[304].

Harvesting in the horizontal wicker hives of Messokampos, Florina, took place every two years, a unique event for beekeeping in Greece. In autumn, during the first cold weather in the region (usually in September), the bee colony gathered in the front half of the wicker hive, where there were dark (i.e. old as long as they were not harvested) honeycombs. During the harvest, the honeycombs of the rear part of the wicker hive, which were one or two years old, were removed. The production of honey ranged between 20 and 30 kg per hive[305].

There is a report of the use of horizontal wicker hives by other refugees coming from Asia Minor in the region of Florina, which were even different from those of Messokampos, but there is no more information[306]. Finally, in the prefecture of Drama (Eastern Macedonia in northern Greece), among other traditional hives, refugees used a cylindrical horizontal open-at-both-ends wicker hive, 100 to 110 cm long, for which, apart from the information of its use in the area, there are no additional data[307].

2
CONSTRUCTION OF WOVEN HIVES

The vast majority of Greek beekeepers themselves wove the wicker hives they used. The purchase of ready-made hives from professional basket weavers was not particularly common, but it was found in several areas. The professional basket weavers who are reported to have also manufactured hives were those of Argos, one of the most important basketry centers in Greece, central and western Crete, Tinos, and the Gypsy basket weavers, who were active throughout the country.

The basket weavers of Argos supplied wicker hives to beekeepers mainly from the eastern Peloponnese, Attica and Boeotia[308]. The beekeepers of these areas used different types of woven hives, and the basket weavers were forced to weave all the types in use to meet the needs of their clientele. These types were the conical skep, the bell-shaped skep, the open-at-the-top wicker hive, another open-at-the-top wicker hive with downwards protruded base, and the open-at-the-top-and-bottom wicker hive. The first two were fixed-comb and the rest movable-comb hives.

For the manufacture of wicker hives, like all other baskets, the basket weavers of Argos used chaste tree (*Vitex agnus-castus*) and reeds. Chaste tree rods were cut from late August to February and stored in covered warehouses, sheltered from exposure to the sun and rain. Ten to fifteen days before using the rods, they were placed in cisterns to moisten and soften. They were then transported to a shady corner of the workshop and covered with a wet cloth to keep them moist. From there, the basket weaver took them and split for the weaving[309].

The reeds were cut in January and February and stacked upright so that rainwater would flow over them and would not make them rot. Before using them, they were cleaned of their outer peel and moistened. They were then split, from four to ten strands, depending on the diameter of each reed and the size of the basket to be woven[310].

For the conical skep, the basket weavers of Argos selected five chaste tree rods for warps, crushed them into two and tied the point of the crease making ten in all (FIG. 60). This was followed by weaving the weft of spit reeds between the ten warps that had been formed (FIG. 61). The weaving continued with reed splits for weft, while when the rods that played the role of warps were covered, others were placed. At two points the weaving with reed splits stopped and chaste tree splits were used for weft, that is, chaste tree belts were created that enhanced the strength of the skep. When the skep reached the desired height, they wove the rim, with the free ends of the warps. Each free warp became a weft and was woven between its upright warps: one from the front and one from behind. The diameter of the mouth was practically calculated with the length of the weaver's hand from the tip of the fingers to the elbow. Thus, the skep had a mouth diameter of about 50 cm, while its height amounted to 60 cm[311].

FIG. 60. Start of construction of a conical skep by a basket weaver from Argos
(Photo: M. Efthymiou, from Efthymiou-Chatzilakou 1979/80).

FIG. 61. Weaving of the upper part of a conical skep by using chaste tree rods as a weft
(Photo: M. Efthymiou, from Efhtymiou-Chatzilakou 1979/80).

The construction of the open-at-the-top wicker hive began with the weaving of a "cross" (FIG. 62) of seven vertical and seven horizontal chaste tree rods. At the junction of the rods of the "cross", which was the central point of the base of the hive, wefts of chaste tree rods were placed before each warp and woven to the right: one warp from the front, two from the rear, i.e. covering a warp from the front and passing the weft behind the next two warps, and so on. When the height of the warps was exhausted, new ones were put into place on which the weaving continued. As the basket weaver wove the base, he constantly pulled the warps outwards, so that the "star" they formed would open and the thick and hard weft could be woven between them. When the base was completed (FIG. 63), the "cross" was turned upside down to continue weaving in the same direction, i.e. from left to right. As soon as the basket weaver reversed the "cross", he tied the rods-warps high up (FIG. 64), cut the rods-wefts that protruded at the level of the base, hit the base with a wooden beater (FIG. 65) to give it a harmonious shape and continued weaving with whole rods for reasons of strength. This part of the body of the hive made of chaste tree rods for weft was called the *stefani* or hoop and was woven one time from the front, two times from behind. The body of the hive was woven with reed splits for weft with intermittent weaving of wefts from chaste tree splits, in order to create zones to enhance its strength. When the weaving reached the desired height, the rim was woven in the manner described above for the skep. The dimensions of the open-at-the-top wicker hive of the basket weavers of Argos were approximately 45 cm height, 40 cm mouth diameter and 26 cm base diameter[312].

FIG. 62. Weaving of the "cross" at the start of open-at-the-top wicker hive construction (Photo: M. Efthymiou, from Efhtymiou-Chatzilakou 1979/80).

FIG. 63. Completion of an open-at-the-top wicker hive base (Photo: M. Efthymiou, from Efhtymiou-Chatzilakou 1979/80).

FIG. 64. Reversal of the "cross" and tying of the warps in an open-at-the-top wicker hive (Photo: M. Efthymiou, from Efhtymiou-Chatzilakou 1979/80).

FIG. 65. Tapping the base of an open-at-the-top wicker hive to achieve a harmonious shape
(Photo: M. Efthymiou, from Efhtymiou-Chatzilakou 1979/80).

The open-at-the-top wicker hive with downwards protruded base was constructed by the basket weavers of Argos a little larger than the open-at-the-top one with a flat base. Its height reached 50 to 55 cm, the diameter of the mouth was 44 to 47 cm and that of the base 28 to 30 cm. For its weaving, 16 warps were used, two more than in the open-at-the-top wicker hives of a flat base. The downwards protruded base of this hive was usually woven with the help of a soup tureen. Otherwise, its weaving was identical to that of other open-at-the-top wicker hives[313].

The bell-shaped skep was identical in size to the open-at-the-top wicker hive with downwards protruded base. The difference in its construction concerned the weaving of the base, which in this case was the roof of the woven hive (since it was placed with the mouth downwards by the beekeepers), was woven initially in the same way as the conical skep and then with the help of a soup tureen. Thus, this skep acquired a protrusion on its roof which, as a handle, facilitated its transportation[314]. Interestingly, only chaste tree rods were used for the weaving of this skep, apparently for greater durability[315].

For the weaving of the open-at-the-top-and-bottom wicker hive, a wooden disc with sixteen holes, in which the sixteen warps were placed, was used. The basket weaver thus wove the body of the basket, and when he reached the desired height proceeded to weave the rim of the mouth. He then removed the wooden disc and wove the now free warps. The height of this skep reached 50 cm, the diameter of the lower mouth 28 cm, and the diameter of the upper mouth 44 cm[316].

In Crete, wicker hives were woven by professionals living in the prefectures of Heraklion, Rethymno and Chania. In the first two, the type of wicker hive that was woven was similar to the *kaniskara*, or a household basket of various uses, but without handles. This wicker hive had 16 warps and for its construction rods from various plants were used, such as lentisk (*Pistacia lentiscus*), chaste tree, and wild olive (*Olea europaea*, subsp. *oleaster*) for the warps, base and rim, and split reeds for weft[317].

Tasos Leontidis, in his study of Cretan baskets mentions that the wicker hive in question was placed upside down, after the beekeeper opened a hole in its roof for the entrance of the bees, over which he placed a ceramic piece for protection against rain[318]. He himself had not seen a wicker hive, placed in the way he describes, since, as he notes, their construction and use had ceased several years before his fieldwork[319]. Such a woven hive, however, has not been mentioned by any of those who have dealt with the traditional beekeeping of Crete, although its production and therefore its use is said – according to Leontidis – to have been remarkable, in the prefectures of Heraklion and Rethymno. Besides, a skep with its entrance in its roof (!) confronts the beekeeping logic. A similar type of hive has not been registered anywhere in the world. I think the author here is wrong, possibly conveying incorrect information given by his basket-weaving informants, who were obviously not familiar with beekeeping. The fact that the drawing of open-at-the-top wicker hive from western Crete cited by Leontidis is attributed to the wrong (upside down) inclination of its side walls (since, the way the wicker hive is sketched, it could not function as a movable-comb hive, as was used by local beekeepers) shows that the author had no knowledge of beekeeping and reinforces my view of error. The woven hive which was manufactured by basket weavers in the prefectures of Heraklion and Rethymno in the shape of the *kaniskara* should have functioned as an open-at-the-top, movable-comb hive. Similar wicker hives are known in Crete (see FIG. 8). These wicker hives are essentially similar in shape to the *vraskia*, the traditional open-at-the-top clay hives that were widely used in the past in these areas[320].

In the prefecture of Chania there were two large basketry centers, Sirili and Pemonia, whose basket weavers also wove hives. In addition to these two centers, there were professional basket weavers in several other villages of this prefecture[321]. The woven hive used here was open-at-the-top-and-bottom, and for its manufacture the basket weavers employed rods of chaste tree or myrtle (*Myrtus communis*)[322] and more rarely from other plants. In some cases, it seems that reed splits were used for weft, though not often, which obviously had to do with their lesser long-term durability[323]. For its weaving (in Sirili), a mold was used, in which the 18 rods constituting the warps of the wicker hive were fastened upright. Around these warps the basket weaver began weaving[324].

Basketry was, and continues to be, one of the traditional professions of Gypsies in Greece. This art, after all, has characteristics that suit nomadic or semi-nomadic populations[325]. Gypsy basket weavers moved around the country weaving and selling their baskets[326]. Among other things, they also made hives if there was an interested clientele. Some Gypsy basket weavers were also permanently established, as in Thrace[327]. Gypsy basket weavers of Thrace used chaste tree and willow (*Salix* spp.) rods as well as reeds. In the central and northern prefecture of Evros, however, the use of cornel (*Cornus mas*) and hazel (*Corylus* spp.) has been recorded, a fact attributed to Bulgarian or more generally Slavic influence[328]. In recent decades, the decline in wild vegetation has led some weavers to resort to the use of materials of lesser strength, such as rods of mulberry (*Morus* spp.), poplar (*Populus* spp.), quince (*Cydonia oblonga*), and even tamarisk (*Tamarix* spp.). These materials are mixed in use with chaste tree rods and unpeeled willow[329].

Andros, which was the only Cycladic island where skeps were used, was visited by wandering basket weavers, mainly from Tinos but also by Gypsies, who wove, among other things, hives for the local beekeepers. The beekeeper informed them about the design he preferred for his skep as well as the size[330].

As was already mentioned, beekeepers usually wove themselves the hives they used. As shown in the earlier chapter, in Thassos, beekeepers bought bee colonies along with their skeps from their colleagues in Chalkidiki. The need for the construction of hives was therefore not particularly great, since after harvesting there was usually a stock of empty skeps[331]. In some cases, however, it was necessary to weave hives on the island. According to the manuscript of Georgios Kourmoulis (dated 1937), for the construction of the local skep, clematis (*Clematis* spp.) rods were used with which they made the roof of the skep as well as the knob of the roof. The rest was woven mainly with chaste tree rods, which were used for warps, as well as weft (the thinnest ones). The skep was supported on four or five legs of parts of branches, being pointed at the upper end (FIG. 66), which were entered into the weave. In the end, they were anointed internally and externally with dung[332].

FIG. 66. One of the legs of a migratory skep from Chalkidiki (Photo: G. Mavrofridis, 2016).

In Chalkidiki, two types of the local skep were manufactured which differed in their size and construction materials. In areas where transport was carried out mainly by pack animals, the smallest type was woven, while in those where transport was carried out by boats, the larger one. My informant, Asterios Giouvannakis, an old beekeeper from Gomati, made a rough demonstration and description of the construction of the smallest type of Chalkidiki skep. According to this, the local skep was woven with 14 warps of chaste tree rods (FIG. 67) from which the knob of the roof was also created. For the weaving of the roof, clematis rods were used. When the roof was completed, with a diameter of about 28 cm, the warps were bent to weave the body of the skep. As a weft clematis rods were used for about 8 cm and then reed splits. In the middle of the skep, a belt with chaste tree splits for weft was created, which imparted durability. The rim at the mouth was woven with chaste tree rods. The legs on which the skep stood were made of hardwood. These legs were placed next to warps within the weave of the skep and protruded from the mouth 5 to 6 cm. Then the skep was anointed inside and outside with a mixture of dung and ashes. The height of the skep reached 40 cm. During weaving, in all the skeps made by the beekeeper, at some point, so as not to be seen easily, a ſtick or marker of other material was incorporated in a particularly peculiar way, as a mark of ownership. In case of theft, the beekeeper was then able to prove that the skep was his.

FIG. 67. Representation of the ſtart of conſtruction of a Chalkidiki skep
(Photo: G. Mavrofridis, 2016).

The smaller type of Chalkidiki skep seems to have had little variations in every region. In the village of Stanos, for example, the skeps were made with 12 warps and not with 14[333]. In some areas, the scorching of the skep has been recorded before its anointing to burn the spikes[334].

In the large type of the Chalkidiki skep, which was mainly manufactured in Kassandra and Mount Athos, for the weaving of the body of the skep, chaſte tree rods were only used, inſtead of reed splits[335].

In Skyros, where the same type of skep with even larger dimensions was found, the height reached 75 cm and the diameter of the mouth 45 cm. For its weaving, chaste tree rods were used for warps (16 of them) and for weft. Clematis rods were used for the knob on the roof. The skep stood on 8 wooden feet[336].

In Thrace, although women also practiced beekeeping, the construction of skeps was considered a male profession[337]. The Thracian beekeepers who wove their hives themselves, used, like the Gypsies of the area, chaste tree, reed or willow[338] and rarely quince rods, as in some cases in the area of Didymoteicho[339]. The Pomaks (Slavic-speaking Muslims) of the region, however, wove their hives using for warps cornel rods, which have great durability over time, and for weft flexible rods from climbing plants or hazel splits[340]. Hazel splits were also used by Pontian Greek basket weavers for the weaving of various types of baskets[341], and the same probably applied to hives. In Koukos, Pieria, however, the skeps woven by its inhabitants of Pontic origin had for weft rods of stranglewort (*Cynanchum acutum*)[342], which may be attributed to the absence of hazel in the area for the supply of raw material.

In the prefecture of Serres, the local skeps seem to have been woven mainly from willow rods which were used both as warps and as wefts[343]. The horizontal wicker hives of Messokambos, Florina were woven exclusively from willow rods. In this case, 13 thick rods, 85 cm long, were used as warps, and thinner willow rods for weft. The warps protruded slightly on either side, while the rims at the edge of the skep were tied with wire[344]. Willow was also used for weaving bell-shaped skeps, the use of which was brought to the region of Florina by refugees from Eastern Thrace. For the construction of these hives, a wheel was used, which was able to turn, like that of a potter. In the center of the wheel the tied edge of the willow warps was placed, while a metal rim held the warps in a fixed position, giving the skep the desired bell-shaped shape. The weaving was carried out with thin willow rods, which were tightly tied on the rim of the skep with other, thinner, rods from the same plant or with thread[345].

In the conical skeps of Florina, cornel or more rarely hazel rods were used as warps, while hazel or willow rods or reed slits were used for weft[346]. In the corresponding skeps of Kastoria, willow, chaste tree, and even honeysuckle (*Lonicera* spp.) was used as a weft[347], while those found in the west of the prefecture of Larissa were woven exclusively with chaste tree rods[348].

In Attica, open-at-the-top wicker hives were woven with chaste tree for the warps, and chaste tree or reed splits for weft[349]. These materials were probably also used for the skeps in the area. For the top-bars of the open-at-the-top wicker hives, some hard wood was used, such as kermes (*Quercus coccifera*), strawberry tree, or wild olive[350].

On the islands of Salamis[351], Aegina[352] and Spetses[353], as warps in the local open-at-the-top wicker hives, they employed chaste tree or lentisk (*Pistacia lentiscus*) rods, and for weft, chaste tree rods or split reeds. In Tracheia, Argolis, chaste tree rods were used for warps, and peeled lentisk at the base for weft, so that it does not rot and can stand the test of time. For weaving the body of the hive, chaste tree rods, reed splits and lentisk rods were used alternately as weft[354].

In Kynouria, woven hives were open-at-the-top-and-bottom and were made with warps made of chaste tree rods, and with weft of split reeds and chaste tree rods. In some cases, however, only chaste tree rods were used as a weft[355]. In Laconia, open-at-the-top-and-bottom local hives were also woven with chaste tree rods and split reeds[356]. The same materials were used for weaving the open-at-the-top and open-at-the-top-and-bottom woven hives on the island of Cythera. For the construction of open-at-the-top-and-bottom hives, a special wooden disc was used, round or oval, with holes through which the warps were inserted to start weaving[357].

Finally, on the island of Andros, the skeps were often woven in a different way by the beekeepers of each village. George Speis has done an excellent job and lists in his monograph on the beekeeping of the island all the versions of the construction of the local hives[358]. Here we will be limited to the most important ones. In the skeps of the island, chaste tree was mainly used for warps (15 to 20), but the use of wild olive, lentisk, and plane tree (*Platanus orientalis*) has also been recorded. For wefts, mainly clematis rods, split reeds and chaste tree rods were used[359].

From the above, it becomes evident that the relatively wide variety in the construction materials of woven hives in Greece is mainly due to the availability of suitable raw materials for weaving in each region. In some cases, beekeeping practices also played a role in the choice of construction materials, which often led to different needs regarding the woven hive used. So, depending on their needs, some beekeepers sought smaller and lighter hives, others larger or/and more durable.

3

WOVEN HIVES IN ANTIQUITY AND THE MIDDLE AGES

Basketry is an ancient art. Its oldest known specimens go back to the Neolithic Age of the Near East and date from the second half of the 10th millennium BC[360]. But when did man start building woven nests for bees? This must evidently have taken place after the beginning of the practice of beekeeping, the oldest testimony of which is considered a fresco discovered in Egypt and dated to the 25th century BC[361]. In Greece, beekeeping seems to have been known during at least the second millennium BC[362], but there is no definite data on the hives used at the time. Unfortunately, ancient Greek literature is not of any help, as no information about the types and materials of constructing ancient hives has been preserved[363].

A number of researchers believe that some pictorial elements from the Bronze Age in Greece represent woven hives. First, Paul Faure, in a letter of his, sent on March 26, 1971 to Vedanabeele and Olivier[364], identified the ideogram *179 of Linear B script, consisting of three horizontal and four vertical lines inclined to the interior in their upper part, with a hive. Apparently, this scholar had in his mind a skep which, to some extent, the said ideogram looks like. Besides, in one of his studies, he believes that the Minoans "placed their bee colonies in skeps or in overturned clay pots with a hole in the base"[365]. His view was based on the discovery of a clay vessel with a hole in the center of its base, in a cave in Agia Paraskevi, Pediada[366], which he considered an open-at-bottom hive.

Later, Manolis Melas, by embracing Faure's opinion, assumes in his turn that similar clay vessels discovered on Crete and Karpathos were used as open-at-the-bottom hives, with the entrance to the center of their base that now becomes a roof. In fact, while searching ethnographic parallels for a similar placement of hives, he finds them in the conical wicker skeps of Eastern Thrace[367], and in the straw skeps of Britain[368]. By going further, he believes that the ideograms of Linear B *134 and *190, "are versions of a basic ideogram (inverted cone with no dashes) which perfectly represents the upside down conical beehive"[369].

The ideograms *179 and *134 / *190 of Linear B script.

Haralambos and Anastasios Harissis accept both Faure's view of the ideogram *179, and that of Melas on the ideograms *134 and *190, believing that they depict woven skeps. In addition, they take symbol 7 of the inscription of the Phaistos Disc (1700 BC) as being a skep[370]. Irene Papageorgiou examined a miniature frieze from Akrotiri of Thera, which, among other things, depicts rows of objects of triangular shape, concluding that a part of the frieze in question concerns an apiary with conical skeps[371]. She also agrees, in her turn, with Faure's view of the ideogram *179 of Linear B, and cites as an ethnographic parallel the conical skeps of Andros, North Macedonia and those of the Pomaks of Thrace, also referring to the use of cylindrical skeps in northern Greece and some islands of the northern Aegean[372]. Finally, she implies that the references of Latin writers to woven hives, which we will examine below, also relate to wicker skeps[373].

The above suggestions for the use of skeps during the Bronze Age in certain areas of southern Greece are mainly based on figurative similarity (ideograms of Linear B, symbol of Linear A or shapes in friezes). The ideogram *179, as well as *134 and *190, which are in fact two versions of the same ideogram[374], do indeed look like conical skeps, but this is not in my view sufficient to identify them with skeps[375]. As for the inscription of the Phaistos Disc, according to the recent proposition by the Welsh linguist, Gareth Owens, the reading is syllabic, and the symbol 7 is said to correspond to the syllable ti[376]. Regarding the ethnographic parallels, these are not related to Crete, the islands of the southern Aegean and the Peloponnese, where wicker skeps were unknown in the local beekeeping tradition.

The wicker skeps of Andros do not represent, in my opinion, the characteristic hive of the beekeeping tradition of the Cyclades, where traditional beekeeping was practiced, until a few decades ago (and in some isolated cases continues to be practiced), with horizontal-type hives, made of clay, stone slabs, and more rarely of woven plant stems. The clay hives of Greek antiquity brought to light by archaeological excavations in the Cycladic islands are also of horizontal type[377] and in some cases probably open-at-the-top[378]. The use of wicker skeps in Andros seems to have been introduced by the Greeks of Eastern Thrace who settled on the southern part of the island in the 14th century[379] and by Arvanites (Christian-Albanians) who settled on the northern part in the 15th century[380]. For this reason, both horizontal types of hives (which continued the Cycladic beekeeping tradition of the local population) and upright open-at-the-bottom or cupboard hives (which were the beekeeping contribution of the foreigners) were found on the island until recently. Over time, there is no evidence, written or otherwise, about the use of traditional skeps in the Cyclades (with the exception of Andros), in the Dodecanese, Crete and the Peloponnese.

As for the view that some of the clay vessels discovered in Crete and Karpathos were upright open-at-the-bottom hives with the entrance to the middle of their roof, lacks beekeeping logic. Besides, nowhere has a similar type of hive been recorded. So, I think that skeps were not in use during the Bronze Age in Greece, just as they were not in Greco-Roman antiquity (which we will deal with below). However, the possibility that other types of woven hives were used at this time cannot be ruled out.

The famous marble omphalos (navel-stone) of Delphi, a copy of the Hellenistic or Roman period, which bears a relief decoration imitating an *agrenon* (ἄγρηνον, fabric of woolen strips), is considered by some to be a sculptural form of skep[381]. The ethnographic parallels offered for this identification concern either straw skeps from northwestern Europe, in this case from France[382], or the bell-shaped traditional wicker skeps of Florina and the similar-shaped wicker skeps of the Rhineland in a depiction from the 15th century[383]. However, straw skeps were throughout the ages completely unknown in Greece. Stalks or stems of straw were not used for the construction of hives in Greece, the Balkans (with the exception of some areas in the interior of Slovenia), Asia Minor and generally around the Mediterranean. The southernmost regions where straw skeps have been recorded are the Alps (in some valleys near the northern border of Italy), the Pyrenees[384] and the aforementioned Slovenia[385]. The use of straw skeps is believed to have originated in the German tribes west of the Elbe River in the last pre-Christian centuries[386], and gradually spread to several areas, mainly due to the movements of the Germanic tribes[387]. As for the wicker skeps of Florina, the origin of which is from Eastern Thrace (as mentioned in the earlier chapter), they do not appear in any source to have been used in ancient Greece.

As already mentioned, the surviving ancient Greek literature does not provide us with data on the types of hives used. In Latin, on the other hand, there is a lot of information, both about the types of hives, and about practicing beekeeping with them[388]. Regarding hives, Varro (Marcus Terentius Varro), who lived between 116 and 27 BC, informs us that the hives made of rods were smeared internally and externally with cow dung and that they were made narrower in the middle to "imitate the shape of the bees" (*Res rusticae*, III.16.15-16). At the ends of these open-at-both-ends horizontal wicker hives were placed lids, while the entrances for the bees, consisting of small holes, were located to the right and left in the middle of the hive. Wicker hives similar to the description of Varro were used during the 20th century in Ethiopia[389].

Virgil (Publius Vergilius Maro), almost fifty years younger than Varro, speaks, among other things, of hives made of woven plant stems, and points out that they should have a narrow entrance to protect against the cold of winter and the heat of summer (*Georgicon*, IV.33-36).

Columella (Lucius Junius Moderatus Columella), who lived in the 1st century AD, had at his disposal the works of Hyginus and Celsus, now lost. He classifies the hives according to their insulating properties, considering the (horizontal) wicker hives the third best, after the ones made of cork and fennel stalks (*Ferula* spp.) (*De re rustica*, IX.6.1).

Pliny the Elder (Gaius Plinius Secundus, 23-79 AD) speaks of horizontal wicker hives with a lid on their back opening, which could be moved to fluctuate their inner space. Otherwise, he also considers the hives of woven flexible branches to be the third best (*Naturalis historia*, XXI.47.80). Pseudo-Quintilian (Pseudo-Quintilianus), about whom we only know that he lived between the 2nd and 4th century AD, refers to the way of construction and coating of woven hives (*Declamationes*, XIII.3). Finally, Palladius (Rutilius Taurus Aemilianus Palladius), a writer of the 4th century AD, simply copies Columella[390].

From the works of Latin writers on the hives of their time it becomes clear that all hives, including of course wicker hives, were of a horizontal type open-at-both-ends. In *Geoponica* also, a compilation of works by older writers that took its final form in the 10th century AD, all the hives described are of a horizontal type[391]. However, there appears the view that in Roman times there were also in use upright open-at-the-bottom hives and in particular skeps[392]. This view is based on a report by Petronius (Gaius Petronius Arbiter, 27-66 AD) (*Satyricon*, 39.14: *Terra mater est in medio quasi ovum corrotundata, et omnia bona in se habet tanquam favus*) and that of Virgil about the existence of woven hives from plant stems that we saw above. What is mentioned by Petronius is understood as a description of an egg-like round hive, which is believed to refer to a skep. However, even if this is accepted, why should an egg-shaped hive be open-at-the-bottom? Egg-shaped skeps cannot be accepted because their underside, the mouth, is flat. On the contrary, a horizontal hive could be egg-shaped. There are also examples of traditional horizontal hives of this shape[393].

Wicker hives, though not skeps, were probably in use in ancient Greece as well. Due to their material of manufacture, however, they were not preserved over time and cannot be discovered by archaeological excavations. Failure to find fragments of clay hives during archaeological investigations in certain areas, contrary to what happens in other areas, where a large number of them are found, is probably due, according to J. L. Davis[394], to the use of wicker hives[395].

Since horizontal[396] and open-at-the-top[397] clay hives were certainly in use in Greek antiquity (the former, from at least the 5th century BC; the latter, from the 3rd century BC at the latest), it is most likely that these types of hives also existed in woven form. The first to connect the ancient Greek movable-comb hives with the traditional movable-comb wicker hives was Abbé Della Rocca of Syros in 1790[398]. He jumped to this view after taking into account the opinion of Angelo Contardi (who fifteen years earlier had referred to the use of movable-comb hives by the ancient Greeks, but without justifying it[399]), which he associated with the use of woven open-at-the-top-and-bottom hives in Crete that he knew from testimonies of third parties. Later, H. M. Fraser, after studying Aristotle's findings on the bee, concluded that these should come from movable-comb hives, similar to those (open-at-the-top wicker hives) encountered in 1676 in Attica by the travelers Spon and Wheler[400]. Fraser's view was shared by many, such as N. Nicolaidis[401],

P. Georgandas[402], P. Papadopoulo[403], F. Ruttner[404], Br. Adam[405], and T. Bikos[406], who agreed with (although independently of) him, that the open-at-the-top wicker hives were used in ancient Greece. The same view was initially held by Eva Crane[407], who, however, later changed her mind, believing that top-bar, movable-comb hives were unknown in Greek antiquity[408].

The earliest references to the use of wicker hives in Greek texts date back to the 4th century AD and come from Saint John Chrysostom (347-407)[409] and Saint Gregory of Nyssa (c.335-c.395)[410]. Hesychius of Alexandria (5th century AD) speaks in his *Lexicon* of a *hive* as a *woven bee vessel*. Later, in the 12th century AD, the *Etymologicon Magnum Lexicon* refers to the hive as a *woven vessel* (549.24). Unfortunately, the above reports do not reveal the type or types of wicker hives.

There is the view that upright hives were generally introduced to the Balkan Peninsula and Greece from the North, by various peoples, such as the Visigoths, the Huns and the Slavs, who invaded Byzantine territory from the 4th century AD onwards[411]. Especially for the skep of Chalkidiki (i.e. the migratory skep of northern Greece) a possible route has been postulated. It begins in the 12th century, at the Serbian monastery (Hilandar Monastery) on Mount Athos, which received as gifts apiaries, both from its founder Stefano Nemanja (1113-1199) and from his successive Serbian kings in the 13th and 14th centuries[412].

As far as upright hives are concerned, these, in their open-at-the-top form, were proven to be in use since at least the Hellenistic period[413], and there is no question of importing vertical hives in general from the North during the Byzantine period. For open-at-the-bottom hives, including skeps, one cannot exclude their origin from the North, though there is no relevant information. In any case, the Visigoths and the Huns cannot have been the bearers of beekeeping with vertical hives in the Balkans[414]. This is because, in that case, we come to the absurd conclusion of attributing to looters and nomads a higher beekeeping culture than a permanently settled population. In other words, we should accept that raiders who attacked the Byzantine Empire for the purpose of plundering, possessed superior beekeeping knowledge and practices which they passed on to the beekeepers of the Empire.

Regarding the skep of Chalkidiki and the possibility of its introduction into the area by the Serbs, we should observe that nowhere is it mentioned that the donations of the Serbian rulers to apiaries (hives) were related to skeps and, in particular, to the migratory skeps of northern Greece (Chalkidiki). After all, beekeeping was already developed in Chalkidiki before the Byzantine Emperor, Alexios III Angelos (1153-1158) granted Hilandar Monastery to Stefan Nemanja and his son Rastko in 1198. As mentioned in Chapter I, a document from mid-10th century contains an order according to which the people of Ierissos should not install apiaries in the area of Mount Athos[415], while it is possible that migratory beekeeping was practiced with the local skeps. Apart from the people of Ierissos, the monks of Mount Athos also practiced beekeeping (before and certainly after the foundation of Hilandar Monastery) and there are several testimonies to this activity[416]. Finally, if the migratory skep of northern Greece had a Serbian origin, there follows that this type of skep should have been used in Serbia as well. However, the Serbian wicker skeps that have been recorded in literature are conical, as are all wicker skeps in the Balkan countries north of Greece[417]. In my opinion, the migratory skep of northern Greece appeared at some point as a result of the evolution of beekeeping at a more or less professional level, along with the full implementation of migratory beekeeping and the consequent further enrichment of the beekeepers' knowledge[418].

In the 12th century, the Jewish rabbi Samuel ben Meir (1085-1158) reported that beekeepers in the "kingdom of Greeks" (that is, in the Greek-speaking Byzantine Empire) managed to produce new bee colonies seven to eight times a year, in contrast to what happened in France, where only three annual reproductions were made[419]. Modern historians, based on this testimony, conclude that "beekeeping is more developed in the Byzantine Empire than in northwestern Europe"[420]. In northern France, however, to which ben Meir refers, it would not be possible, regardless of beekeeping knowledge and skills, to multiply bee colonies seven or eight times, due to climatic conditions (low temperatures occurring much earlier in autumn, longer winter duration, low temperatures in spring, etc.) and the characteristics of the local subspecies of bee, *Apis mellifera mellifera*, which do not help in this direction[421].

The multiplication of bee colonies seven to eight times a year is also difficult in some areas of Greece. Bueras, the Athenian informant of the traveler John Hawkins, who was already mentioned, reports at the end of the 18th century that in Athens a bee colony could multiply two or even three times in excellent years, while the resulting bee colonies were also amenable to separation[422]. This means that, under favorable conditions, the bee colony could give three more colonies and the new colonies another three, that is, a total of six bee colonies, in addition to the maternal one. However, the great multiplication of bee colonies would certainly be at the expense of honey production. For this reason, according to Bueras and the abbot of the Penteli Monastery, the beekeepers of Attica destroyed the queen cells or killed the queens of the swarms, so as not to weaken the hives. Bueras and the beekeepers of Athens were happy if they simply doubled their hives within a year[423]. It is likely, therefore, that the maximum possible increase in bee colonies per year should have been carried out mainly in cases where a beekeeper sought an increase in his hives for productive beekeeping the following year or perhaps for sale to other beekeepers.

Another area where a multiplication of bee colonies has been reported at such levels is Chalkidiki. There, the beekeepers who used the local skep, while practicing beekeeping to sell bee colonies to their colleagues, managed to multiply the hives six or, under favorable conditions, eight times. The earliest testimonies for the practice of this reproductive beekeeping in Chalkidiki date back to the middle of the 19th century[424].

The combination of the above information with the testimony of ben Meir, allows the possibility that the practice of reproductive beekeeping, regardless of whether or not there was a sale of bee colonies to other beekeepers, was applied in some areas of Greece from the 12th century onwards.

4

THE INFLUENCE OF GREEK WOVEN HIVES ON THE EVOLUTION OF WORLD BEEKEEPING

The most important Greek traditional hives were undoubtedly the movable-comb hives, and this is because of their mode of operation that allows a series of beekeeping manipulations, which are either impossible to carry out with other types of hives or are carried out with greater difficulty. As mentioned in Chapter I, these hives, in the form of open-at-the-top wicker hives, were encountered, in 1676, at the Monastery of Kaisariani on Mount Hymettus, by the travelers Jacob Spon and George Wheler, who described them in their works, thus communicating their existence to English and French-speaking readers. However, while Spon, who was mainly interested in antiquities, made a simple description (without though avoiding some mistakes)[425], Wheler devoted two pages of his book to the wicker hives of Hymettus and the way they work, even citing a relevant drawing (see FIG. 5). Wheler's account of the way the monks of the monastery multiplied and harvested their hives made an impression in England, where at the same time (and for another 200 years) it was commonplace to kill bees for the harvest. Thus, Wheler's information on the "Greek hives" offered the inaugural impetus for the creation of hives that would allow the raising, one by one, of the honeycombs and consequently the manipulations of the Greek monks[426].

However, although Wheler's description was informative and detailed, comprehensible even to someone unfamiliar with beekeeping, his drawing for the wicker top-bar hive of Hymettus contradicted both his writings (!) and the traditional wicker hives of the area, as we have known them from later descriptions. Eva Crane was the first to note this mismatch[427]. Specifically, although the text refers to wicker hives, the drawing depicts attributes of a hive made of straw strips, a practice known at the time of the author in central and northwestern Europe, including England, but completely unknown in Greece[428]. Furthermore, top bars, although described as broad and flat, which even need a knife to be separated, are depicted in the drawing extremely thin and at a distance from each other.

Next to these two mismatches, however, further inconsistencies are encountered. The text states that the hive was smeared with mud and dung, an element that is absent from the drawing[429]. The proportions of the wicker hive are also differentiated from those we know from the traditional wicker top-bar hives of Attica (see FIG. 26). The installation of the hive on a wooden raised base was probably not known. A similar practice was in use in Britain, but it has not been recorded in Attica or elsewhere in Greece. Finally, the entrance of the bees, as depicted in the drawing, does not correspond to the entrance we know from the traditional hives[430].

Here we are dealing with a paradox. The drawing that essentially changed the way beekeeping was practiced worldwide did not correspond to reality, even though the effort in the Western world to practice mobile-comb beekeeping was mainly based on this drawing. So, if a beekeeper, having seen the drawing in question, tried to build a similar movable-comb hive, it must be considered certain that the result would be disheartening. The thin top-bars with space between them would urge the bees to attach their combs vertically to them and no movable combs would be created. This comes from my own experience, when, during the practice of experimental beekeeping with movable-comb hives, in the absence of a sufficient number of top bars of normal width, I used noticeably thinner ones, leaving space between them[431].

The desired result could perhaps only be reached by carefully reading Wheler's text and by not taking into account the drawing (though, usually, the opposite is the case). However, this alone would not be enough because an important parameter is missing, that of the width of the top-bars. Wheler himself did not understand the importance of the correct width of the top-bars and does not cite this extremely important information. So, even if someone (based on the text) placed the top-bars next to each other without making them have the correct width, the result would still be negative.

It is therefore concluded that, only after (enough) experimentation with them, could Wheler's information regarding the creation of movable-comb hives have a practical effect. This is the main reason, in my opinion, why the open-at-the-top wicker hive did not spread in Europe immediately and widely, in its well-known traditional (Greek) form, but only later and in an "improved" form, a result of the work by various researchers, who made the creation of the desired movable-comb possible. Typical are the examples of the open-at-the-top woven hives constructed in 1768 by Thomas Wildman[432] (FIG. 68) and in 1847 by Robert Golding[433] (FIG. 69), which, however, also leave space between the top-bars.

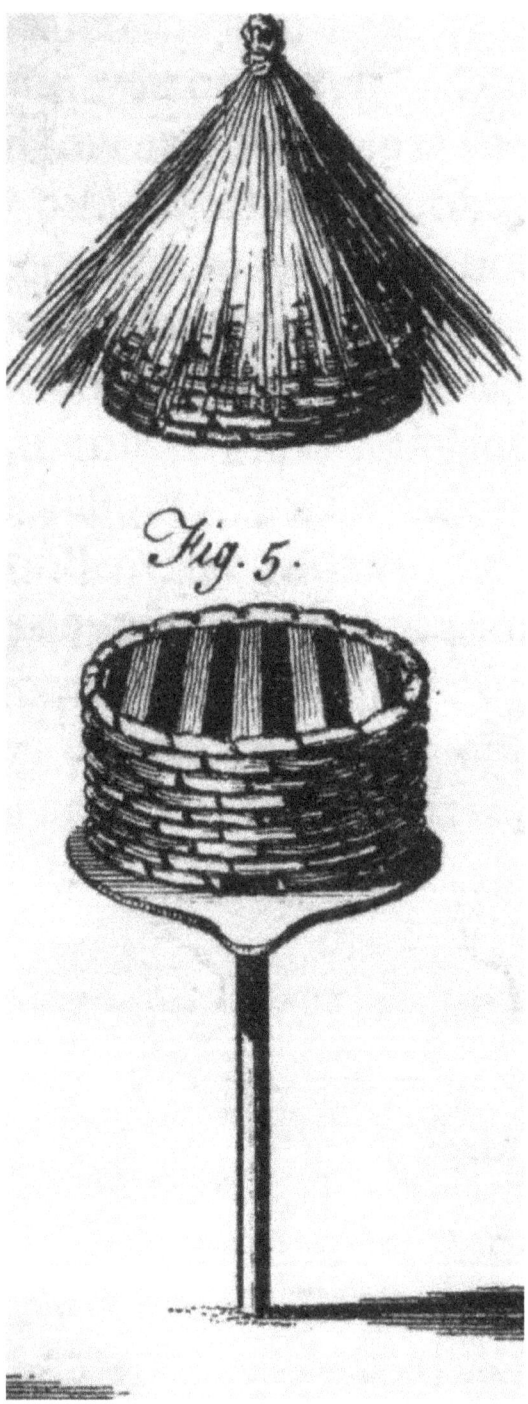

FIG. 68. Woven hive created under the influence of the Greek movable-comb wicker hive in 1768 by Th. Wildman (1770, pl. 2, fig. 7).

FIG. 69. The "improved Grecian hive" by R. Golding (1847, 25).

The failure to understand the importance of the width for the top-bars by Spon and Wheler, combined with the erroneous drawing cited in the latter's work, seems to have had a negative effect on the spread of the idea of the movable comb and significantly delayed (by almost two centuries according to Eva Crane[434]) the evolution of world beekeeping.

Worth-mentioning here are two more descriptions of traditional top-bar, movable-comb hives of Greece, although they did not affect the evolution of world beekeeping, since they were not published in their time. The first belongs to the Cretan Zuanne Papadopoli who, in 1696, wrote (in Italian) in Padua, memoirs of his life in Venetian-occupied Crete. Papadopoli was himself a beekeeper on the island in the 1630s, a fact that gives special weight to his writings related to beekeeping[435]. The hives of Crete, as he describes them[436], fully correspond to the traditional *vraskia*, or clay hives of similar shape to open-at-the-top wicker hives, as they have been known to us since the 20th century[437]. Papadopoli was the only one who referred to the width of the top-bars of the traditional movable-comb hives. He describes them as "two fingers in breadth", that is, as the sum of the width of the index finger and the middle finger, a method of practical calculation with which most of the Greek users of these hives, empirically determined the "bee space". However, this valuable information of Papadopoli remained unexploited, since his work was not published until for three centuries.

The other description of traditional movable-comb hives, the wicker hives of Attica, was carried out by the traveler John Hawkins and is based on the information he collected from the abbot of Penteli Monastery in Attica and by an elderly Athenian beekeeper around 1796. Hawkins also cites a drawing of a wicker hive from Athens (see FIG. 7) with its dimensions, and mentions the number of top-bars that were placed on it[438]. He had not realized the great importance of the width of top-bars and did not wish to know more about it. However, since he mentions the diameter of the mouth of the hive and the number of bars, the width could be deduced, with a little effort and experimentation, by any restless beekeeper. Unfortunately, Hawkins' text was not published until half a century later, just a few years before Lorenzo L. Langstroth identified the "bee space" and created the modern movable-frame hive.

So, after the reports of the first travelers about the Greek open-at-the-top wicker hive, there were several others who, in their effort for a more rationalized and effective beekeeping practice, used the idea of movable comb to build a more suitable hive for this purpose. Some, such as the aforementioned T. Wildman and R. Golding, actually tried to copy the Greek top-bar wicker hive. Others, such as Abbé Della Rocca from Syros, who knew from eyewitness descriptions the movable-comb wicker hives of Crete, built board hives with top-bars. So, Della Rocca himself proposed a two-storey hive with two rows of top-bars, which were not placed next to each other, but had a small gap in-between[439]. Later, Robert Kerr, in 1819, proposed a complex octagonal hive, T. M. Howatson, in 1827, came up with a hive consisting of two wooden boxes, and Johannes Dzierzon, in 1848, one that featured one or more series of movable combs. Other researchers, such as an unidentified British author with the initials "J. A."[440], in 1683 (i.e. just one year after the publication of Wheler's book in England) and Huber, Playfair, Prokopovich, Munn, Dubeauvoys and Shaw, all focused their efforts on the construction of hives based on movable-frames[441].

This is in general terms the evolution of the idea of movable comb, which led to the creation, by L. L. Langstroth, in 1851, of the modern movable-frame hive and to the definition of the "bee space". Langstroth's relation with earlier efforts and the Greek movable-comb wicker hives is confirmed from the books found in his library[442], namely those by Wildman (with his proposal for a woven hive created under the influence of the Greek movable-comb wicker hive) and Golding (with his proposal for an "improved Grecian hive", that is, movable-comb woven hive)[443].

When, with the wide spread of the movable-frame hive around the world – in the second half of the 19th and especially in the first half of the 20th century – it was found that this hive was not suitable for beekeeping in the developing world and especially in sub-Saharan Africa, due to the prohibitive cost for the local population (both of the hive, as well as the necessary equipment of its associated beekeeping practice), the solution was provided by the Greek movable-comb wicker hives. Poppy Papadopoulo who was a beekeeping expert at the Ministry of Agriculture of Greece[444], took over in the 1960s the Department of Beekeeping at the Ministry of Agriculture of Rhodesia (present-day Zimbabwe) and realizing the impossibility of using the frame hive in the area, proposed the use of top-bar wicker hives (she knew from her homeland), which could easily and at no cost be constructed by the local population[445]. Because of their movable combs, these wicker hives were offered for beekeeping practice with lots of interventions on the part of the beekeeper, in contrast to the local traditional horizontal hives of fixed comb that were used there. She then asked the Head of the Department of Beekeeping of the Greek Ministry of Agriculture, Panagiotis Georgantas, to send her an open-at-the-top wicker hive to serve as a model. Georgantas sent her one from the island of Salamis[446] (FIG. 70) and the results were astonishing.

The Greek wicker hive, in its open-at-the-top and a little later also in its open-at-the-top-and-bottom form, began to be widely used in the area[447]. The local sub-species of bees (*Apis mellifera scutellata*) is extremely aggressive, and with modern movable-frame hives most of the bee colony is able to exit (from the gaps between the frames) immediately the hive is opened and defend it by attacking, thus complicating the work of the beekeeper. In the open-at-the-top wicker hives, however, and more generally in the movable-comb hives, the top-bars of which are firmly attached to each other, only a small part of the population exits the hive, i.e. mainly the bees that are located in the honeycombs of the bars that are raised, and in the corresponding sides of the adjacent ones, greatly facilitating the work of the beekeeper.

Papadopoulo's idea of using woven movable-comb hives was followed along the way by others in Africa, such as J. Nightingale in Kenya, who experimented with a modified form of the "Greek hive" (as the Greek top-bar wicker hive became widely known), J. Linder in Senegal, who introduced the hive "David" which was also based on the "Greek hive"[448], and R. D. Guy, who proposed, and practiced beekeeping himself, using woven hives, but with a rectangular mouth, so that their top-bars could be of equal length and could be used in any location of every hive[449]. Ole Hertz later proposed the construction of cheap movable-comb woven hives with the use of baskets used in Gambia to transport the various fruits[450]. The same hive was mentioned shortly after by Sam Manga, who gave technical advice for its construction[451]. Finally, one of the hives used nowadays in Senegal is called "improved Greek hive"[452], and comes from the open-at-the-top wicker hive.

FIG. 70. Movable-comb wicker hive from Salamis sent as a sample to Rhodesia. It is an exhibit at the "Museum of Beekeeping" of Zimbabwe (Photo: P. Papadopoulo, 1960s).

After the success of Poppy Papadopoulo's venture with the introduction of the Greek movable comb hive for rationalized beekeeping in Africa, many began to propose various types of rectangular movable-comb hives (so that the top-bars could be of the same length) from several other materials. The University of Guelph, Canada, proposed for East Africa a board hive with oblique, inward converging sides, dubbed Kenya Top Bar Hive. The creation of this hive, which was destined to become very popular, was attributed to Professor G. F. Townsend and a group of his Kenyan students[453]. A little earlier, in 1965, C. J. Tredweel and P. Paterson (a student from Kenya) had built a similar hive at Sparsholt College of Agriculture, UK[454]. At some point, there arose disagreements, about who was the one who first "discovered" the rectangular movable-comb hive, that is, whether it was the English college or the University of Guelph. Similar hives, however, were in use traditionally in Greece long before their "discovery" by the mentioned researchers[455]. N. Nicolaidis had also written, ten years earlier, in 1955, in the international journal *Bee World*, about the rectangular movable-comb hives traditionally used on the island of Cythera[456].

In summary, the traditional wicker movable-comb hives of Greece have had a dual influence on world beekeeping. As soon as they became known in the West, from travelers' descriptions, they triggered the search for a more rational way of practicing beekeeping. This quest resulted in the creation of the modern movable-frame hive that is predominantly used today in the developed world. When it turned out that this modern hive was not suitable for the developing world, it was proposed to use movable-comb hives, initially copies of the traditional wicker hives of Greece. Later it was proposed to use wooden or made of other materials hives with a rectangular mouth, but still similar to traditional hives of Greece. Nowadays movable-comb hives are widely spread, not only in sub-Saharan Africa, for which they were originally proposed, but also in the southern part of Asia, as well as in Central America.

EPILOGUE

Greek traditional beekeeping possesses a large number of types of hives that include all the main types of traditional hives, i.e. upright open-at-the-bottom, upright open-at-the-top, horizontal open-at-one-end, and open-at-both-ends, as well as some peculiar types, such as wall hives, cupboard hives, built hives, and hives carved into rock. For the construction of these hives, various materials were used from those that were available in each area, such as plant stems, boards, clay, hollowed tree trunks, stone slabs, porous stone, and, in some rare cases, even animal skin. An important role among these materials was played by the various flexible rods, with which the corresponding hives were woven. Thus, over time, a wide variety of types and sub-types of woven hives was created, covering the needs of beekeepers in every region. It is the data that I tried to gather and thereby, in a sense, to "rescue" these hives, through the writing and publishing of this book.

The main objective of this research was to present a more or less complete picture of the Greek woven hives. This includes their use over the centuries, the way and materials of their manufacture, their types and sub-types, the methods and practices of beekeepers who used them, as well as the influence that some of them exerted on the evolution of world beekeeping. I believe that this objective has, more or less, been achieved. On the other hand, I understand that by being the first such attempt it may have some weaknesses or certain points that require further investigation. Besides, the ambition of this work was not to exhaust the issue of Greek woven hives, although such an effort was undertaken, but to make it known and attractive to the wider public.

I hope that the book will be of interest not only to specialists, archaeologists, ethnographers, beekeeping scientists but also to beekeepers who are interested in the history of their art. Finally, it would be desirable if the book were able to trigger the interest of other researchers to deal with issues of traditional beekeeping, about which there are still living informants, at least in the countries around the Mediterranean, either practitioners or eye-witnesses.

NOTES 1 – WICKER-HIVE TYPES AND THE PRACTICE OF BEEKEEPING

1.1.1 Open-at-the-top and open-at-the-top-and-bottom wicker hives

1 Bee space, which was first precisely identified by L. L. Langstroth in 1851, is the distance that must exist between two honeycombs in order for bees to circulate between them. When this interval is shorter, the bees shorten it; if it is longer, they build in-between new honeycomb.

2 Zymbragoudakis 1979, 135; Bikos 2012, 242.

3 Mavrofridis 2007c, 134; 2008a, 168; 2010e, 429. This same measure for calculating the width of top-bars is mentioned in the 17th century by Zuanne Papadopoli (*L'occio* 133r).

4 Mavrofridis 2013b, 25; 2017a, 300; 2018c, 68, 78, note 2.

5 Mavrofridis 2015a, 178-179; 1917a, 302-304, 306.

6 Spon 1678, II, 224-225.

7 Simopoulos 1990, I, 698; Vigopoulou 2005, 7-8.

8 Mavofridis 2010b, 106-107; 2017a, 309.

9 Wheler 1682, 412-413.

10 Cotton 1842, 106a; Mavrofridis 2010e, 431; 2012a, 175; 2017a, 307 and note 57; Mavrofridis & Chairetakis 2019, 42.

11 Thompson 1744, I, 352-353.

12 Spencer 1973, 153; Mitsi 2003, 64; Khatib 2003, 195.

13 Mavrofridis 2010b, 109.

14 Ibid., 109.

15 Della Rocca 1790, II, 466.

16 Ibid., 465-473 and 498-500.

17 Galt 1812, 382.

18 Mavrofridis 2010b, 110; 2017a, 304.

19 Cotton 1842, 107a.

20 The woodcut is not found in the original text by Sibthorp (in R. Walpole 1817, 149 & 1818, 2nd ed., 149) from which Cotton drew. Cotton's previous text, which ends at the page with the woodcut, is by Karl Gustav Fielder (1791-1853) about the honey of Hymettus,

21	Opening from above traditional hives that in Attica (from at least the 17th century onwards) were exclusively made of wicker.
22	Mavrofridis 2012b, 403.
23	Cotton 1842, 103a-106b. Hawkins had passed on his manuscript text in 1802 to Davies Gilbert (1767-1839), the future president of the Royal Society, who seems to have handed it over, in his turn, in 1838, to Cotton who included it in his book.
24	Kambouroglou 1889, 317-318.
25	Mavrofridis 2012b, 400; 2017a, 312. Besides, the same abbot is said to have given information, in 1794, about the livestock farming of Attica, to another traveler, the well-known English botanist, John Sibthorp, who appears as Hawkins's companion at the time. See Walpole 1817, 141.
26	Cotton 1842, 103a.
27	Ibid., 103a-103b.
28	Perhaps under the word "wasp" here is also meant the hornet *Vespa orientalis* which is the greatest enemy of bees in the southern part of central Greece, as well as in southern Greece. In Greek the word "sfika" (σφήκα) is used for both wasp and hornet.
29	Cotton 1842, 103a-103b.
30	Pallis 2009, 259-262.
31	Beaujour 1800, I, 167-168.
32	Mavrofridis 2019a, 122-123.
33	This refers to the well-known, in Attica, surname Bouras, coming from "bouri" which in Albanian (Arvanitian) means "man" (the information is owed to Andromachi Economou, senior researcher of the Hellenic Folklore Research Centre, Academy of Athens).
34	The Arvanites, especially the men, also used to speek Greek.
35	Cotton 1842, 104a-105b.
36	Ibid., 104a. For Buera's reference and its possible meaning, see Mavrofridis 2012b, 401.
37	Cotton 1842, 105b-106a.
38	He refers to fifteen days instead of the correct sixteen, but this can be explained.
39	This method was also known at that time by the monks of the monastery of Penteli. See the above account of their abbot.
40	This is correct, but there are currently methods that allow the bee colonies to join together. Of the traditional beekeepers of Greece, only the people of Thassos, Rhodes and partly of Andros knew and applied these methods
41	Cotton 1842, 104a-105a. See also Mavrofridis 2014c, 413.

42   He rather refers to the greater moth (*Galleria mellonella*), though we cannot exclude the possibility that it could be the lesser moth (*Archoia grisella*). Besides, he does not describe them, but only refers to a "worm" enemy of the bees in the bad years, during which the beehives are weak.

43   From this description, it becomes clear that he refers to *Braula coeca* louse.

44   See note 28.

45   Cotton 1842, 104a.

46   Ibid., 104b-105a.

47   Cotton 1842, 105b.

48   Ibid., 106a.

49   Mavrofridis 2014c, 414.

50   Benton 1894: 723.

51   Frantzeskanis 1903, 7; Georgandas 1957, 287; Zymbargoudakis 1979, 136-137; Ruttner 1979a, 223 and 117, Abb. 1; 1979b, 7-12; Mavrogenis 1979a, 21; Crane 1983, 197-199; 1999, 398-400; Br. Adam 1983, 76, 79-80; Savvakis 1994, 179-180; Bikos 1996a, 267-269; 2012, 243-246; Rammou & Bikos 2000, 429-430; Bikos & Rammou 2000, 23-24; 2002, 11; Mavrofridis 2007c, 137; 2012a, 177; 2017a, 301-302; 2019d, 4-6. Also the manuscripts of Antonioudakis 1964 and Perakis 1973.

52   Mavrogenis 1979a, 21.

53   Antonioudakis 1964, ms.

54   Leontidis 1986, 40, 88-89, 97-98, 102-103; Savvakis 1994, 179.

55   Mavrofridis 2016b, 199.

56   Ibid., 196. For the archaeological findings concerning the possible use of bee gardens in Crete in antiquity, see Price and Nixon 2005, 674-675; Francis 2016, 93-94; Nixon and Moody 2017, 490.

57   Zymbargoudakis 1979, 137.

58   Frantzeskakis 1903, 8; Zymbargoudakis 1979, 137.

59   Frantzeskakis 1903, 8; Loukaki 1964, ms; Ruttner 1979a, 221.

60   Savvakis 1994, 177, who etymologizes the word from katina (backbone). The same word was used by the beekeepers of Arachnaio, Argolis. The name cantinelles, however, is also used in the Italian text by Zuanne Papadopoli (Papadopoli *L'occio* 133r-133v) in the 17th century.

61   Frantzeskakis 1903, 8.

62   Frantzeskakis1903, 8; Bikos 1996a, 268-269; 2012, 246.

63   Papavasileiou 2008, 116.

64   Frantzeskakis 1903, 46-47; Mavrogenis 1979a, 21.

65   Chestnut, in addition to nectar, offers honeydew derived from insects that parasitize on it.

66   Mavrofridis 2020, 405; Mavrofridis et al. forthcoming.
67   Mavrofridis 2020, 405; Mavrofridis et al. 2021; Mavrofridis et al., forthcoming.
68   Mavrogenis 1979a, 21.
69   Zymbargoudakis 1979, 137; Bikos 2012, 243.
70   Loukaki 1964, ms; Zymbargoudakis 1979, 137. Similar smokers seem to have been in use on the island since the Minoan Age. See Tyree et al. 2012, 223-224.
71   Loukaki 1964, ms; Pelekanakis 1964, ms; Perakis 1973, ms; Zymbargoudakis 1979, 137.
72   Mavrogenis 1979b, 245.
73   Frantzeskakis 1903, 66; Zymbargoudakis 1979, 137; Sellianakis 1998, 33-34, 36.
74   Zymbargoudakis 1979, 137; Sellianakis 1998, 34.
75   Della Rocca 1790, II, 496-97 and III, pl. IV, fig. 4, 6.
76   Ibid., II, 497.
77   Bikos 1995a, 14.
78   Ibid., 13-15.
79   Ibid., 15-16.
80   Mavrofridis 2012a, 177; 2017a, 303, note 27.
81   Bikos 1995b, 175-176.
82   Ibid., 176-177.
83   Bikos 1995a, 11, 14.
84   Bikos 1995b, 176.
85   Which according to P. Dimitropoulos (1982, 111) was "very rare".
86   Bikos 2011a, 32.
87   Bikos 2004, 90-91.
88   Roumeliotis 1993, ms; Anonymous 1998, 10; Mavrofridis 2012a, 176.
89   Anonymous 1998, 11-12; Mavrofridis 2012a, 176-177.
90   Roumeliotis 1993, ms.
91   Idid.
92   Ibid.
93   Mavrofridis 2012a, 176; 2013b, 19-20.
94   Typaldos-Xydias 1981, 12.
95   Mavrofridis 2010d, 350; 2010e, 429; 2017a, 305.
96   Mavrofridis 2010d, 349.
97   Mavrofridis 2010d, 349; 2015a, 179.
98   Mavrofridis 2010d, 349-350; 1015a, 179.
99   Br. Adam 1983, 80.

100 Crane 1983, 15; 1999, 316.
101 Bikos 2010a, 324-326.
102 Mavrofridis & Petanidou 2022b, 400-402.
103 Mavrofridis 2010e, 429-430; 2016b, 198.
104 Mavrofridis 2010e, 431.
105 Mavrofridis 2008a, 168-169.
106 Ibid., 169.
107 Today it forms an exhibit of the museum collection at the Institute of Agricultural Sciences.
108 Mavrofridis 2008a, 167-168.
109 Bikos 1995d, 423-424.
110 Bikos 1995c, 383-387.
111 Mavrofridis & Chairetakis 2019, 43.
112 Bikos 1995d, 424.
113 Mavrofridis 2007c, 138 where an older bibliography is gathered. For more recent information, see Bikos 2009c, 334-335; Mavrofridis 2012a, 175-177; 2017a, 308.
114 Crane 1983, 199; Bikos 1998, 538; Rammou & Bikos 2000, 431; Mavrofridis 2018c, 69.
115 See Efthymiou-Chatzilakou 1979/80, 73, fig. 33; 1981/82, 12, fig. 2.
116 Bikos 2009c, 335.
117 Bikos 1999, 6-7; Bikos & Rammou 2000, 19-20.

1.1.2.1 The migratory skep of northern Greece

118 Mavrofridis 2011b, 212; 2017b, 41-42.
119 Mavrofridisς 2017b, 41.
120 Papachrysanthou 1975, 196.
121 Ibid., 201-202.
122 Coloumella (1st century AD) mentions transfer of bees from Achaia to Attica and Euboea, and from the Cyclades to Skyros (*De re rustica* IX.14.19-20).
123 At the end of the 19th and the beginning of the 20th century, beekeepers of Ierissos and other areas of Chalkidiki transported about ten thousand skeps to Mount Athos. See Smyrnakis 1903, 499; Typaldos-Xydias 1927, 13.
124 Tsiriktzidou 1989, 127-128.
125 The oka was an Ottoman unit of weight.
126 Beaujour 1800, I, 255.
127 Mavrofridis 2017b, 43.
128 Urquhart 1938, II, 135-136.

129 Papaoikonomou 2012, 17; 2014, 12.
130 Eckert 1943, 6.
131 Conze 1860, 26-17. However, he does not describe the skeps he mentions.
132 About the author behind the initials W.G.C. see Carty 2000, 353.
133 Karastergios & Kokkora 2010, 11.
134 Clark 1862/63, 307.
135 Mavrofridis 2016c, 11; 2017c, 286.
136 Newpaper *Hermes* (Ερμής), February 20, 1873.
137 Tozer 1890, 296.
138 Mavrofridis 2017c, 274; 2017f, 434.
139 Typaldos-Xydias 1927, 13, 33; Stojianov1942, 101; Eckert 1943, 5; Bikos 2009a, 20; Mavrofridis et al., forthcoming.
140 Br. Adam 1983, 77, 142-143.
141 Kourmoulis 1937, ms; Stojianov 1942, 101.
142 Typaldos-Xydias 1927, 33.
143 Idid., 30.
144 Topalidis1940, 160; Mavrofridis et al., forthcoming.
145 Kourmoulis 1937, ms.
146 On Thassos, as in several other areas of Greece, the queen was formerly considered a male insect. This misconception goes back to antiquity. Aristotle, first, in the *Historia animalium*, in several places, refers to the "king" or "ruler" of bees.
147 Kourmoulis 1937, ms.
148 The method of joining bee colonies with the use of water, flour or bran was also known by the traditional beekeepers of Rhodes, but it concerned the horizontal types of hives they treated (see Vrontis 1938/48, 204).
149 Typaldos-Xydias 1927, 25.
150 Topalidis1940, 160.
151 Typaldos-Xydias 1927, 36.
152 Mavrofridis 2015b, 12; 2016a, 36-37.
153 Dermatopoulos 1954, 65; 1984, 107.
154 Typaldos-Xydias 1927, 27.
155 Mavrofridis 2009c 396-399; 2016a, 37-38.
156 This text, from the newspaper *Echos of Arnea* (Αντίλαλοι της Αρναίας) 1976, is quoted verbatim by D. Kyrou in two of his studies: Kyrou 2000, 372 and 2005, 406.
157 Shinas 1887, 564-565.
158 The article was located and partially republished by Kyrou 2005, 407. For the conclusions

derived from this text see Mavrofridis 2016a, 37-38 and 2017c, 277.
159   Typaldos-Xydias 1927, 29-30.
160   Mavrofridis 2016a, 37; 2017c, 281-282, 287.
161   Mavrofridis 2017c, 282.
162   Typaldos-Xydias 1927, 12-13, 19; Eckert 1943, 5; Mavrofridis et al., forthcoming.
163   Tsellios 1988, 175. According to 1939 data, the families of 150 professional beekeepers of Arnaia owned from one to five mules, depending on the number of their hives. See Papaioannou 1939, 155-184 and Gardikas 2015, 175.
164   Typaldos-Xydias 1927, 16-17; Mavrofridis et al., forthcoming.
165   Typaldos-Xydias 1927, 16-17.
166   Typaldos-Xydias 1927, 13.
167   Testimony by Asterios Giouvannakis (1932-2019) from Gomati, Chalkidiki in 2016.
168   Typaldos-Xydias 1927, 13-15; Mavrofridis et al., forthcoming.
169   According to my informant, Asterios Giouvannakis. The opinion that in the pine forests of Thassos the production was greater than in the corresponding ones of Chalkidiki is also shared by Typaldos-Xydias 1981, 10.
170   Typaldos-Xydias 1927, 13.
171   Smyrnakis 1903, 499; Papaioannou 1939, 178; Mavrofridis et al. 2022a, 25.
172   Typaldos-Xydias 1927, 21-22.
173   Typaldos-Xydias 1927, 26; Tsellios 1996, 186; Yfantidis 1997, 535; Anonymous 2003, 48; Katsaros 2011, 30.
174   See related photo in Typaldos-Xydias 1927, 20.
175   Kourmoulis 1937, ms. See also Mavrofridis 2017c, 295.
176   Typaldos-Xydias 1927, 20-21.
177   Ibid.
178   Dermatopoulos 1984, 107; Karakasis 1994, 96-97.
179   Testimony by Asterios Giouvannakis, the last old beekeeper of Gomati, who sold bee colonies to his Thassian colleagues.
180   Typaldos-Xydias 1927, 33; Mavrofridis et al., forthcoming.
181   Bikos 2008c, 224; Mavrofridis et al., forthcoming.
182   Liakos 2006, 263; Mavrofridis et al., forthcoming.
183   Liakos 2006, 261-262; Bikos 2008c, 226-228.
184   Bikos 2008c, 224-225.
185   Liakos 2006, 261-262; Bikos 2008c, 226-228
186   Typaldos-Xydias 1927, 22-23 (for Chalkidiki); Liakos 2006, 262 (for Skyros); Mavrofridis 2017c, 294-295 (for Thassos, based on information from the Kourmoulis manuscript).

187 Mavrofridis 2017d, 125-126; 2017h, 76-77; 2021b, 6-8.
188 Tsilogeorgis 2011, 703-704.
189 Papaggelos 2000, 201.
190 Typaldos-Xydias 1927, 25.
191 Liakos 2006, 261 fig. 4, 263; Bikos 2008c, 229 fig. 27.
192 Mavrofridis 2008b, 322. A skep broadly similar in shape to the migratory one of northern Greece, but without a woven knob on its roof, was also used in Gonnoi, at the southern foothills of Lower Olympus (Kouvounas 2004, 223-224). A similar skep is exhibited at the Folklore Museum of Gonnoi.
193 Kouvounas 2004, 222-223.
194 Kouvounas 2021, 252.
195 Mavrofridis 2019c, 160.
196 Typaldos-Xydias 2017, 29, 39.
197 My informant, Christos Toskas from Agios Dimitrios, Argolis, had bought and used skeps himself from Chalkidiki in the 1980s.
198 See Crane 1999, 219, fig. 23.4.a, where a photo of a Cretan beekeeper using skeps of Chalkidiki.

1.1.2.2 The migratory skep of Attica and Boeotia

199 Toufexis 1904, 85; 1909, 76.
200 Typaldos-Xydias 1927, 40-41, 46.
201 Stouraitis 1907, 3.
202 Typaldos-Xydias 1927, 42-44.
203 Typaldos-Xydias 1927, 40-44; Mavrofridis et al. forthcoming.
204 Anonymous 1907, 4.
205 Typaldos-Xydias 1927, 43.
206 Mavrofridis 2019a, 124.
207 Typaldos-Xydias 1927, 41, 43, 49. Among the wild plants of the area of Lake Copais, the most important was the conium (*Conium maculatum*).
208 Typaldos-Xydias 1927, 41, 43; Mavrofridis et al. forthcoming.
209 Typaldos-Xydias 1927, 46.
210 Typaldos-Xydias 1927, 41; Mavrofridis et al. forthcoming.
211 Andromachi Economou, senior researcher at the Hellenic Folklore Research Centre of the Academy of Athens, who has long dealt with the economy of the communities of the region (including Kriekouki and Vilia), found no evidence of the existence of professional beekeepers (oral communication on 8/11/2017).

212 Angelos Typaldos-Xydias (1927, 51) states that the income of the migratory beekeepers of Attica and Boeotia is obtained "by saving the required time, without neglecting their usual crops".

1.1.2.3 Conical and bell-shaped skeps

213 Vakarelski 1977, 155-157; Bojiadzhiev 1983, 53; Crane 1999, 220-222; Liapis 2007, 15, 30, 67-68.
214 Eleftheriadis 1992, 26-27; Mavrofridis 2014a, 186-187.
215 Mavrofridis 2008b, 320.
216 Mavrofridis & Tselios 2013, 89.
217 Crane 1999, 176; 2003, 243.
218 Crane 2003, 244.
219 Bodenheimer 1942, 16, 44 fig. IX, A.
220 Br. Adam 1964, 79.
221 Br. Adam 1977, 59.
222 Bikos 2005, 212.
223 Katsouleas 2000, 357.
224 Bodenheimer 1942, 14. However, he does not satisfactorily justify this position and does not mention the period, which he believes the use of conical skeps in Asia Minor dates back to. It seems that he accepts as an axiom that traditional Asia Minor hives were exclusively horizontal, and therefore, any upright type should be a foreign influence on the region.
225 Ibid., 13.
226 Hatzopoulos 1977, 62-63.
227 Anagnostopoulos 1996, 6; Bikos 1997b, 227-228; Rammou & Bikos 2000, 434.
228 Raichevski 1996, 81.
229 Primovski 1960, 648; 1973, 347; Vakarelski 1977, 155-156; Bojiadzhiev 1983, 53; Raichevski 1996, 81.
230 Bikos 2014a, 111-112.
231 Mavrofridis 2007b, 79-80; Bikos 2014a, 111-113.
232 Anonymous 1969, 233. Skeps of this type are exhibited in the Folklore Museum of Didymoteicho.
233 Alexiadis 2006, 90; Liapis 2007, 46, 116, 121; Mavrofridis 2009b, 346; Bikos 2014b, 206.
234 Papadopoulos 2010.
235 Efstratiou 1984, 26.
236 Bikos 2014a, 112.

237 The Hartlib Papers. Letter to S. Hartlib (16 Dec. 1659). Ref: 32/1/1A-5B: 1B, 2A. (www.hrionline.ac.uk/hartlib).
238 Papadopoulos 2010; Bikos 2014a, 112.
239 Crane 1999, 251-256.
240 Gouridis 1967, 235.
241 Anonymous 1969, 233.
242 Bikos 2009c, 336.
243 Mavrofridis and Tselios 2013, 87-89.
244 Mavrofridis 2008b, 320.
245 Mavrofridis 2008b, 320; Bikos 2009b, 117.
246 Bikos 2005, 212, 215.
247 Anagnostopoulos 1996, 6; Bikos 1997b, 226-227; Rammou & Bikos 2000, 434.
248 According to my informant, Konstantinos Hatzis (b. 1921), an old beekeeper of Achladochori from whom I received information in 2013.
249 Mavrofridis & Tselios 2013, 87-88.
250 Bikos 2005, 213-214.
251 Bikos 1997b, 226-227.
252 Kouvounas 2000, 399; 2004, 222; 2005, 168-169.
253 Bikos 2011c, 394.
254 Bikos 1994, 356-358; Rammou & Bikos 2000, 431-432.
255 Pallis 2019, 88.
256 Bikos 2014c, 322; Houliaras 2003.
257 Loukopoulos 1938, 389.
258 Bikos 1996b, 314; Bikos & Rammou 2000, 16; 2002, 7; Speis 2016, 36.
259 Speis 2016, 42.
260 Mavrofridis & Petanidou 2022a, 273.
261 Bikos 2011b, 251; Speis 2016, 37, 47.
262 Speis 2016, 36-37, 45.
263 Bikos 1996b, 315-317.
264 Speis 2016, 175; Mavrofridis & Petanidou 2022a, 273-274. The practice of reducing the entrance of hives for protection from Vespa orientalis is a long tradition in the Cyclades. Clay lids of horizontal hives with many small holes – bee entrances, dating to the Hellenistic period, have been revealed in Paros, testifying that the same problem was faced by ancient beekeepers. See Mavrofridis 2015c, 89-91; 2018a, 849-853.
265 Speis 2016, 176.
266 Ibid., 48.

267 Ibid., 137, 146.
268 Ibid., 158.
269 Testimony of my informant Miltiadis Zouras (b. 1948), from Kato Varidi, Andros in 2020.
270 Mavrofridis 2007a, 24.
271 Speis 2016, 158.
272 Mavrofridis 2019c, 159.
273 Mavrofridis 2013a, 34.
274 Crane 1999, 239-240; Mavrofridis 2013c, 160.
275 Gouridis 1967, 235; Hatzopoulos 1977, 64; Anagnostopoulos 1996, 10; Bikos 1997a, 121.
276 Anagnostopoulos 1996, 10.
277 Houliaras 2003; Bikos 2014d, 441-442.
278 Kouvounas 2000, 399.
279 Speis 2016, 147-148.
280 Mavrofridis & Tselios 2013, 88.
281 Bikos 2005, 215.
282 Bikos 1994, 356-357.
283 Mavrofridis & Tselios 2013, 88-89.
284 Bikos 2010b, 403.

1.2 Horizontal Wicker Hives

285 Bikos 2009d, 415.
286 Bikos 2008b, 78.
287 Mavrofridis 2007d, 162.
288 Bikos & Rammou 2000, 24; 2002, 11.
289 Bikos 2015, 251.
290 Bikos 2008b, 78.
291 Bikos 2006a, 10; 2006b, 100.
292 Mavrofridis 2010c, 278.
293 Bikos 2006a, 10-12.
294 Fiorini 1880, 376.
295 Rizopoulou-Igoumenidou 2000, 399.
296 Melanofrydis 1955, 3035; Mavrofridis 2010c, 278.
297 Bodenheimer 1942, 13; Mavrofridis 2014b, 261.
298 Bikos 1997b, 228-229; Anagnostopoulos 2000, 305-306.

299 Anagnostopoulos 2000, 307.
300 Bikos 1997b, 228; Anagnostopoulos 2000, 308.
301 Anagnostopoulos 2000, 308.
302 Pattinson 2012, 349.
303 Ibid.
304 Mavrofridis 2021a, 69.
305 Anagnostopoulos 2000, 307.
306 Ibid., 309.
307 Bikos 2009c, 336; 2014a, 111.

NOTES 2 – CONSTRUCTION OF WOVEN HIVES

308   Efthymiou-Chatzilakou 1979/80, 64.
309   Ibid., 58-59.
310   Ibid., 58.
311   Ibid., 60-62, 71.
312   Ibid., 60-62, 66, 72.
313   Ibid., 66, 72.
314   Ibid., 72.
315   On the other hand, chaste tree rods gave weight to the skep.
316   Efthymiou-Chatzilakou 1979/80, 66, 72.
317   Leontidis 1986, 40, 88-80, 97-99 and 145-162.
318   Ibid., 40, 89, 98.
319   Ibid.
320   Mavrofridis 2019, 2-4. The earliest evidence of the use of *vraskia* dates back to the 1630s and is owed to Papadopoli (*L'occio*, 133r-133v).
321   Leontidis 1986, 102-104.
322   Mavrogenis 1979a, 21.
323   Perakis 1973, ms; Leontidis 1986, 40.
324   Leontidis 1986, 103.
325   Liapis 2007, 36.
326   Liapis 2007, 39-46; see also Stathaki-Koumari 1985, 20.
327   Liapis 2007, 37.
328   Ibid., 37-39.
329   Ibid., 37.
330   Speis 2016, 44.
331   Mavrofridis 2017c, 288.
332   Kormoulis 1937, ms.
333   Anonymous 2003, 47.
334   Tsellios 1996, 185-186.
335   Typaldos-Xydias 1927, 20-21.
336   Bikos 2008c, 224.
337   Papadopoulos 2010; Bikos 2014a, 112.

338 Gouridis 1967, 235; Liapis 2007, 37, 47; Mavrofridis 2007b, 79-80; Papadopoulos 2010; Bikos 2014a, 112; 2014b, 206.

339 Oral testimony by Mrs. Chrysoula Tsakiraki-Kyroudi from the Folklore Museum of Didymoteicho in 2019.

340 Liapis 2007, 46, 121.

341 Ibid., 47.

342 According to my informant Ioannis Nikolaidis (b. 1925) in 2007.

343 According to my informants, Charalambos Konstantinidis (b. 1925), from Angistro, in 2007 and Konstantinos Chatzis (b. 1921), from Achladochori, in 2013.

344 Anagnostopoulos 2000, 305.

345 Bikos 1997b, 227-228.

346 Anagnostopoulos 1996, 5-6; Bikos 1997b, 226-227.

347 Bikos 2005, 212.

348 Kouvounas 2000, 399.

349 Georgandas 1967, 287; Mavrofridis 2012a, 174.

350 Bikos 2009c, 335; Nicolaidis 1955, 145; Georgandas 1967, 287.

351 Bikos 1995d, 423.

352 According to my informant from Mesagros, Panagiotis Chaldaios (b. 1920) in 2008.

353 According to my informant, Panagiotis Mathios (b. 1948) in 2008.

354 Mavrofridis 2010d, 350.

355 In the open-at-the-top-and-bottom hives of Kynouria that are exhibited in the museum collection of the Institute of Agricultural Sciences, reed splits or chaste tree rods were used as weft.

356 Roumeliotis 1973, ms; Anonymous 1998, 10.

357 Bikos 1995a, 15-16.

358 Speis 2016, 34-47.

359 Speis 2016, 34-36.

360 Belogianni 1991, 66

NOTES 3 – WOVEN HIVES IN ANTIQUITY AND THE MIDDLE AGES

361 For the wall paintings depicting beekeeping scenes of Egypt, see Crane 1999, 161-167; Kriśtky 2010, 11-14; Mavrofridis 2011a, 52-54.
362 Mavrofridis 2006, 268-269; Harissis & Harissis 2009, 6-17; Harissis 2018a, 79-88.
363 Mavrofridis 2022, 1.
364 For this letter see Vendenabeele & Olivier 1979, 287.
365 Faure 1973, 159.
366 Savvakis 1994, 175; Mavrofridis 2006, 269.
367 Refers to the work of Hatzopoulos (1977) on the Kryonero of Eastern Thrace.
368 Melas 1999, 488.
369 Ibid., 489.
370 Harissis & Harissis 2009, 25. See also Harissis 2018b, 25-26 and Harissis 2020, 27.
371 Papageorgiou 2016, 101-114.
372 Ibid., 104-106.
373 Ibid., 104-105.
374 The ideogram *190 is considered to represent the (now former) ideogram *134. See Melena 2014, 140.
375 According to José Melena (2014, 140) the ideogram in question probably means "salt and brine (when liquid)".
376 Owens 2018.
377 See Mavrofridis 2022, 3 where the relevant bibliography is collected.
378 Mavrofridis 2018b, 108; Zachos 2021, 255-256
379 Mavrofridis et al. 2022b, 633
380 Polemis 1981, 102; Giohalas 2010, 17-18.
381 Richards-Mantzoulinou 1979, 72-73, 8; Harissis & Harissis 2009, 24-25.
382 Richards-Mantzoulinou 1979, 72-73, 8; Harissis & Harissis 2009, 24-25.
383 Harissis & Harissis 2009, 24.
384 Mavrofridis 2018d, 416.
385 Juvanec 2002, 3-10.
386 Fraser 1950, 33-34; 1955, 177; 1958, 12, 17, 20. The use of strips to make baskets was however known from the Neolithic Age in the Lake Constance area (see Andonova et al.

387 Fraser 1955, 177-178; 1958, 12, 17, 20; Crane 1999, 251-252.
388 For details on the beekeeping information of the Latin authors see Fraser 1951, 29-39, 40-80, 100-101; Crane 1994, 118-132; Mavrofridis 2011c, 266-269; 2018b, 102-105.
389 Crane 1994, 121.
390 Whitfield 1956, 114.
391 *Geoponica*, XV. See Mavrofridis & Goutzamani 2019.
392 Harissis & Harissis, 2009, 25; Harissis, 2020, 27.
393 Crane 1983, 51.
394 Davis 1996, 462.
395 However, it could also be due to the use of hives made from other perishable materials, such as hollow logs or boards, which Davis does not mention.
396 Rottrof 2006, 126-127, for the hitherto known bibliography, and Mavrofridis 2022, 3-4, for the more recent one.
397 See Andesron-Stojanovic & Jones 2002, 349-351, 355-365; Mavrofridis 2013b, 17-27; 2022, 5-6, 9-12.
398 Della Rocca 179, II, 465-466, 498.
399 Contardi 1775, 16-17, note 12.
400 Fraser 1951, 17-18, 94.
401 Nicolaidis 1955, 145.
402 Georgandas 1957, 286.
403 Papadopoulo 1965, 4.
404 Ruttner 1979a, 229.
405 Br. Adam, 1983, 79.
406 Bikos 2000, 285-287.
407 Crane 1977, 184.
408 Crane 1983, 201-202; 1999, 398.
409 *Patrologia Graeca*, 62.105.
410 *Patrologia Graeca*, 46.657.
411 Crane 1999, 219.
412 Ibid.
413 Andesron-Stojanovic & Jones 2002, 349-351, 355-365; Mavrofridis 2013b, 17-27; 2022, 5-6, 9-12.
414 Mavrofridis 2011b, 209; 2016c, 10; 2017c, 275.
415 Papachrysantou 1975, 196.
416 See Papaggelos 2000, 190.

417 Mavrofridis 2011b, 21; 2016c, 10; 2017c, 276.
418 See also Chapter 1.1.2.1 (*The migratory skep of northern Greece*).
419 Kazhdan & Epstein 2009, 62; Krauss 2018, 113.
420 Anagnostakis 2000, 178; 2016, 11.
421 Mavrofridis 2017g, 460.
422 It should not escape us here that in Athens there were used movable-comb wicker hives, which were easily divided in order to multiply, without their user being forced to hunt for swarms or to carry out the time-consuming relevant manipulations of beekeepers who used the migratory skep of northern Greece or Attica.
423 See Chapter 1.1.1 (Open-at-the-top and open-at-the-top-and-bottom movable-comb wicker hives).
424 See Chapter 1.1.2.1 (*The migratory skep of northern Greece*).
425 On this issue see Chapter 1.1.1 (Open-at-the-top and open-at-the-top-and-bottom movable-comb wicker hives).

NOTES 4 - THE INFLUENCE OF GREEK WOVEN HIVES ON THE EVOLUTION OF WORLD BEEKEEPING

426   Mavrofridis 2017e, 189-190; 2019b, 3-4.

427   Crane 1999, 395.

428   Mavrofridis 2018d, 416.

429   The straw skeps of central and northwestern Europe were not smeared with mud or dug either externally or internally.

430   Mavrofridis 2010b, 107-108; 2017a, 310.

431   Mavrofridis 2017a, 310; 2017e, 190.

432   Wildman 1770, Pl. 2, fig. 7.

433   Golding 1848, 25-26.

434   Crane 1997, 5.

435   Harissis & Mavrofridis 2012a, 57; 2012b, 271-272.

436   Papadopoli *L'occio*, 133r-133v.

437   Mavrofridis 2019d, 4; 2020, 397-398.

438   Cotton 1842, 105b-106a.

439   Della Rocca 1790, II, 467-469.

440   J. A. may have been John Aubrey (1626-1697). See Crane 1999, 414.

441   Crane 1999, 414-421; Kritsky 2010, 106-110; Marcenay 1979, 56.

442   Johansson & Johansson 1972, 22-27; Crane 1983, 211; 1999, 422.

443   Mavrofridis 2017e, 191-192.

444   In the 1930s, P. Papadopoulo worked for the spread of the modern movable-frame hive among the beekeepers of Central Greece and Crete.

445   Bikos 2008a, 23-24.

446   Bikos 1993, 297. The wicker hive was sent by Georgandas "a few years after the Apimondia Congress in Vienna" which was held in 1956 (in my estimate in the early 1960s).

447   Papadopoulo 1965, 4-5; 1967, 3-5; 1969, 18.

448   Crane 1971, 34.

449 Guy 1971, 18-24.
450 Hertz 1994, 8-9.
451 Manga 1995, 5.
452 Hussein 2001, 46.
453 Kigatiira 1976, 10-13; Crane 1999, 423.
454 Crane 1983, 212; 1999, 423; Paterson 2008, 97.
455 Mavrofridis 2009a, 288-290; Mavrofridis & Anagnostopoulos 2012, 483-484.
456 Nicolaidis 1955, 146.

BIBLIOGRAPHY

Primary Sources

Antonioudakis 1964 (manuscript): Αντωνιουδάκης, Π. Λαογραφικά εκ Κρήτης. Χειρόγραφο αρ. 2844. Αρχείο Χειρογράφων του Κέντρου Ερεύνης της Ελληνικής Λαογραφίας της Ακαδημίας Αθηνών.

Varro, *Res Rusticae*: Varron, *Èconomie rurale* (ed. C. Guiraud), Livre III. Paris 1997 [Les Belles Lettres].

Virgil, *Georgicon:* Vergilius, *Georgicon (*ed. J. B. Greenough). Boston 1900 (Ginn & co).

Geoponica: *Geoponica sive Casiani Bassi scholastici de re rustica eclogae* (ed. H. Beckh). Lipsiae 1895 [Teubner].

Gregory of Nyssa: Γρηγόριος Νύσσης, Λόγος Γ΄: Εἰς τὸ ἅγιον Πάσχα, καὶ περὶ τῆς ἀναστάσεως ἐλέχθη τῇ μεγάλῃ Κυριακῇ (ed. J. P. Migne). *Patrologiae cursus completus. Series Graeca*, Tomus XLVI. Paris 1863.

Hesychius, *Lexicon*: *Hesychii Alexandrini Lexicon* (ed. K. Latte), II (E-O). Copenhagen 1966.

John Chrysostom: Ιωάννης Χρυσόστομος, 'Ομιλία ΙΕ΄: Ὑπόμνημα εἰς τὴν πρὸς Ἐφεσίους ἐπιστολήν (ed. J. P. Migne). *Patrologiae cursus completus. Series Graeca*, Tomus LXI. Paris 1862.

Columella, *De re rustica*: Columella, *De re rustica* (ed. E. S. Foster, E. Heffner), Vol. 2. London 1960 [Loeb].

Pseudo-Quintilian, *Declamationes*: *Declamationes XIX maiores Quintiliano falso acriptae* (ed. L. Håkanson). Stuttgart 1982 [Teubner].

Kourmoulis 1937 (manuscript): Κουρμούλης, Γ. Ι. Γλωσσική ύλη εκ Θάσου. Χειρόγραφο αρ. 588. Αρχείο Χειρογράφων του Κέντρου Ερεύνης των Νεοελληνικών Διαλέκτων και Ιδιωμάτων της Ακαδημίας Αθηνών.

Loukaki 1964 (manuscript): Λουκάκη, Μ. Λαογραφικά Κρήτης. Χειρόγραφο αρ. 2839. Αρχείο Χειρογράφων του Κέντρου Ερεύνης της Ελληνικής Λαογραφίας της Ακαδημίας Αθηνών.

Etymologicon Magnum: *Etymologicon Magnum* (ed. Th. Gaisford). Oxford 1848 (Typographeo Academico).

Palladius, *Opus agriculturae*: Palladius, *Traité d'agriculture*, Livre I et II (ed. R. Martin). Paris 1976 [Les Belles Lettres].

Papachrysanthou 1975: *Actes du Prôtaton* (ed. D. Papachrysanthou). Paris 1975 (Archives de l'Athos VII).

Papadopoli, *L'occio*: Papadopoli Zuanne. *L'occio (Time of Leisure). Memories of seventeenth-century Crete* (ed. A. Vincent). Venice 2007 (Hellenic Institute of Byzantine and Post-Byzantine Studies).

Perakis 1973 (manuscript): Περάκης, Γ. Λαογραφικά στοιχεία Μουστάκου Χανίων. Χειρόγραφο αρ. 3676. Αρχείο Χειρογράφων του Κέντρου Ερεύνης της Ελληνικής Λαογραφίας της Ακαδημίας Αθηνών.

Pelekanakis 1964 (manuscript): Πελεκανάκης, Ε. Συλλογή λαογραφικής ύλης. Χειρόγραφο αρ. 2854. Αρχείο Χειρογράφων του Κέντρου Ερεύνης της Ελληνικής Λαογραφίας της Ακαδημίας Αθηνών.

Petronius, *Satyricon*: Petronius Arbiter (ed. M. Haseltine). London 1913 [Loeb].

Pliny the Elder, *Naturalis Historia*: Pline l'Ancien, Histoire naturelle (ed. J. André), Livre XXI. Paris 1969 [Les Belles Lettres].

Roumeliotis 1993 (manuscript): Ρουμελιώτης, Π. Γ. Συλλογή λαογραφικού υλικού από την περιοχή Ζάρακα της Επαρχίας Επιδαύρου Λιμηράς του Νομού Λακωνίας. Χειρόγραφο αρ. 4500. Αρχείο Χειρογράφων του Κέντρου Ερεύνης της Ελληνικής Λαογραφίας της Ακαδημίας Αθηνών.

Secondary Sources

Alexandrou 1985: Αλεξάνδρου, Α. *Πετρώνιου Σατυρικόν* (μετάφραση, πρόλογος, σχολιασμός). Αθήνα (Νεφέλη).

Alexiadis 2006: Αλεξιάδης, Δ. Η μεγάλη οικογένεια των μελισσοκόμων. *Μελισσοκομική Επιθεώρηση*, 20(2): 90.

Anagnostakis 2000: Αναγνωστάκης, Η. Βυζαντινή μελωνυμία και μελίκρατος πότος. Αντιλήψεις για τη χρήση των μελισσοκομικών προϊόντων στο Βυζάντιο ως τον 11° αιώνα. In: *ΣΤ' Τριήμερο Εργασίας, Η μέλισσα και τα προϊόντα της*, Νικήτη 12-15 Σεπτ. 1996. Αθήνα (ΠΤΙ ΕΤΒΑ), pp. 161-189.

Anagnostakis 2016: Αναγνωστάκης, Η. Πρόλογος στο Γερμανίδου Σ., *Βυζαντινός μελίρρυτος πολιτισμός*, Αθήνα (ΕΙΕ), pp. 11-12.

Anagnostopoulos 1996: Αναγνωστόπουλος, Ι. Θ. *Μελισσοκομία και μελισσοκομική παράδοση στην περιοχή της Φλώρινας*. Αθήνα (Μουσείο Μπενάκη).

Anagnostopoulos 2000: Αναγνωστόπουλος, Ι. Θ. Δύο τύποι εγχώριων κυψελών της φλωρινιώτικης μελισσοκομίας. In: *ΣΤ' Τριήμερο Εργασίας, Η μέλισσα και τα προϊόντα της*, Νικήτη 12-15 Σεπτ. 1996. Αθήνα (ΠΤΙ ΕΤΒΑ), pp. 295-310.

Anderson-Stojanovic & Jones 2002: Anderson-Stojanovic, V. R., Jones, J. E. Ancient beehives from Isthmia. *Hesperia*, 71(4): 345-376.

Andonova et al. 2020: Andonova, M., Million, S., Stelzmer, I. Archaeobotany of basketry: Identification of Neolithic coiled baskets from the pile dwelling settlements of Lake Constance, Germany. *Integrated Microscopy Approaches in Archaeobotany*. Reading.

Anonymous 1841: Anonymous. Fielder's journey through every part of the Kingdom of Greece. *The Foreign Quarterly Review*, XXVI, pp. 337-369.

Anonymous 1907: Ανώνυμος. Ο τρυγητός των μελισσιών κατά το παρελθόν έτος. *Μελισσοκομική Εφημερίς*, 2(6): 4.

Anonymous 1969: Ανώνυμος. Λαογραφικά Κουφόβουνου Διδυμοτείχου. *Θρακικά*, 43: 222-257.

Anonymous 1998: Ανώνυμος. *Κυψέλες της Ελλάδας*. Αθήνα (ΕΔΙΑΜ).

Anonymous 2003: Ανώνυμος. Τα κοφίνια. *Μελισσοκομικό Βήμα*, 1(4): 47-48.

Balogianni 1991: Μπελογιάννη, Μ. Καλαθοπλεκτική και ψαθοπλεκτική: τέχνες με παρελθόν. *Αρχαιολογία*, 40: 64-69.

Beaujour 1800: Beaujour, F. *Tableu du commerce de la Grèce*. II T., A Paris (Ant. – Aug. Renouard).

Benton 1894: Benton, F. Apiculture in Germany. *American Bee Journal*, 34(23): 723.

Bikos 1993: Μπίκος, Θ., Μελισσοκομικές καταγραφές. *Μελισσοκομική Επιθεώρηση*, 7(9): 295-297.

Bikos 1994: Μπίκος, Θ. Μελισσοκομικές καταγραφές. *Μελισσοκομική Επιθεώρηση*, 8(10): 353-358.

Bikos 1995a: Μπίκος, Θ. Μελισσοκομικές καταγραφές. *Μελισσοκομική Επιθεώρηση*, 9(1): 11-16.

Bikos 1995b: Μπίκος, Θ. Μελισσοκομικές καταγραφές. *Μελισσοκομική Επιθεώρηση*, 9(5): 175-179.

Bikos 1995c: Μπίκος, Θ. Μελισσοκομικές καταγραφές. *Μελισσοκομική Επιθεώρηση*, 9(11): 383-387.

Bikos 1995d: Μπίκος, Θ. Μελισσοκομικές καταγραφές. *Μελισσοκομική Επιθεώρηση*, 9(12): 423-426.

Bikos 1996a: Μπίκος, Θ. Μελισσοκομικές καταγραφές. *Μελισσοκομική Επιθεώρηση*, 10(7-8): 267-272.

Bikos 1996b: Μπίκος, Θ. Μελισσοκομικές καταγραφές. *Μελισσοκομική Επιθεώρηση*, 10(9): 314-320.

Bikos 1997a: Μπίκος, Θ. Μελισσοκομικές καταγραφές. *Μελισσοκομική Επιθεώρηση*, 11(3): 120-123.

Bikos 1997b: Μπίκος, Θ. Μελισσοκομικές καταγραφές. *Μελισσοκομική Επιθεώρηση*, 11(5): 226-232.

Bikos 1998: Μπίκος, Θ. Μελισσοκομικές καταγραφές. *Μελισσοκομική Επιθεώρηση*, 12(12): 536-541.

Bikos 1999: Μπίκος, Θ. Μελισσοκομικές καταγραφές. *Μελισσοκομική Επιθεώρηση*, 13(1): 5-8.

Bikos 2000: Μπίκος, Θ. Κινητή κηρήθρα. In: *ΣΤ' Τριήμερο Εργασίας, Η μέλισσα και τα προϊόντα της*, Νικήτη 12-15 Σεπτ. 1996. Αθήνα (ΠΤΙ ΕΤΒΑ), pp. 284-288.

Bikos 2004: Μπίκος, Θ. Μελισσοκομικές καταγραφές. *Μελισσοκομική Επιθεώρηση*, 18(2): 87-92.

Bikos 2005: Μπίκος, Θ. Μελισσοκομικές καταγραφές. *Μελισσοκομική Επιθεώρηση*, 19(4): 212-216.

Bikos 2006a: Μπίκος, Θ. Μελισσοκομικές καταγραφές. *Μελισσοκομική Επιθεώρηση*, 20(1): 9-13.

Bikos 2006b: Μπίκος, Θ. Μελισσοκομικές καταγραφές. *Μελισσοκομική Επιθεώρηση*, 20(2): 96-100.

Bikos 2007: Μπίκος, Θ. Μελισσοκομικές καταγραφές. *Μελισσοκομική Επιθεώρηση*, 21(1): 31-35.

Bikos 2008a: Μπίκος, Θ. Μελισσοκομικές καταγραφές. *Μελισσοκομική Επιθεώρηση*, 22(1): 22-26.

Bikos 2008b: Μπίκος, Θ, Μελισσοκομικές καταγραφές. *Μελισσοκομική Επιθεώρηση*, 22(2): 78-84.

Bikos 2008c: Μπίκος, Θ., Μελισσοκομικές καταγραφές. *Μελισσοκομική Επιθεώρηση*, 22(4): 223-229.

Bikos 2008d: Μπίκος, Θ. Μελισσοκομικές καταγραφές. *Μελισσοκομική Επιθεώρηση*, 22(5): 365-371.

Bikos 2009a: Μπίκος, Θ. Μελισσοκομικές καταγραφές. *Μελισσοκομική Επιθεώρηση*, 23(1): 14-20.

Bikos 2009b: Μπίκος, Θ. Μελισσοκομικές καταγραφές. *Μελισσοκομική Επιθεώρηση*, 23(2): 112-117.

Bikos 2009c: Μπίκος, Θ. Μελισσοκομικές καταγραφές. *Μελισσοκομική Επιθεώρηση*, 23(5): 334-338.

Bikos 2009d: Μπίκος, Θ. Μελισσοκομικές καταγραφές. *Μελισσοκομική Επιθεώρηση*, 23(6): 412-417.

Bikos 2010a: Μπίκος, Θ. Μελισσοκομικές καταγραφές. *Μελισσοκομική Επιθεώρηση*, 24(5): 322-326.

Bikos 2010b: Μπίκος, Θ, Μελισσοκομικές καταγραφές. *Μελισσοκομική Επιθεώρηση*, 24(6): 398-404.

Bikos 2011a: Μπίκος, Θ. Μελισσοκομικές καταγραφές. *Μελισσοκομική Επιθεώρηση*, 25(1): 32-37.

Bikos 2011b: Μπίκος, Θ. Μελισσοκομικές καταγραφές. *Μελισσοκομική Επιθεώρηση*, 25(4): 250-254. .

Bikos 2011c: Μπίκος, Θ. Μελισσοκομικές καταγραφές. *Μελισσοκομική Επιθεώρηση*, 25(6): 394-400.

Bikos 2012: Μπίκος, Θ. Μελισσοκομικές καταγραφές. *Μελισσοκομική Επιθεώρηση*, 26(4): 240-246.

Bikos 2014a: Μπίκος, Θ. Μελισσοκομικές καταγραφές. *Μελισσοκομική Επιθεώρηση*, 28(2): 110-116.

Bikos 2014b: Μπίκος, Θ. Μελισσοκομικές καταγραφές. *Μελισσοκομική Επιθεώρηση*, 28(3): 205-209.

Bikos 2014c: Μπίκος, Θ. Μελισσοκομικές καταγραφές. *Μελισσοκομική Επιθεώρηση*, 28(5): 322-326.

Bikos 2014d: Μπίκος, Θ. Μελισσοκομικές καταγραφές. *Μελισσοκομική Επιθεώρηση*, 28(6): 440-444.

Bikos 2015: Μπίκος, Θ. Μελισσοκομικές καταγραφές. *Μελισσοκομική Επιθεώρηση*, 29(242): 249-259.

Bikos & Rammou 2000: Bikos, Th., Rammou, E. Les ruches de la Mer Egée. In: *4me Colloque sur l'Apiculture Traditionnelle, Miel. Abeilles et Pierres*, Fontan 24-25 Juin 2000. Fontan (Musée des Arts et Traditions Apicoles), pp. 9-28.

Bikos & Rammou 2002: Bikos, Th., Rammou, E. Beehives of the Aegean islands. *Bee World*, 83(1): 5-13.

Bodenheimer 1942: Bodenheimer, F. S. *Studies on the honey bee and beekeeping in Turkey*. Istanbul (Merket Ziraat Mücadele Enstitüsü).

Bojiadzhiev 1983: Бояджиев, Р. Пчеларство и бубарство. In: *Етнография на България*. Т. II, София (БАН), pp. 53-57.

Br. Adam 1964: Brother Adam. In search of the best strains of bees: Concluding Journeys. *Bee World*, 45(2): 70-83.

Br. Adam 1977: Brother Adam. In search of the best strains of bees: Supplementary journey to Asia Minor. *Bee World*, 58(2): 57-66.

Br. Adam 1983: Brother Adam. *In search of the best strains of bees*. Hamilton (Dadant & Sons).

Carty 2000: Carty, T. J. *A dictionary of literary pseudonyms in the English language*. New York and London 2000 (Routledge).

Clark 1862/1863: Clark, W. G. From Athos to Salonica. *Macmillan's Magazine*, T. VII, Nov. 1862 – Apr. 1863, p. 307.

Contardi 1775: Contardi, A. *Guida sicura pel governo delle api, di Daniele Wildman, colle annotazioni di Angelo Contardi*. Cremona (L. Manini).

Conze 1860: Conze, A. *Reise auf den Inseln des Thrakischen Meers*. Hannover 1860 (Carl Rümper).

Cotton 1842: Cotton, W. Ch. *My bee book*. London (Rivingston).

Crane 1971: Crane, E. Frameless movable-comb hives in beekeeping development programmes. *Bee World*, 52(1): 33-37.

Crane 1977: Crane, E. Beehives, bees and beekeepers. In: *Proceedings from XXVI International Apicultural Congress*. Adelaide (Apimondia), pp. 183-189.

Crane 1983: Crane, E. *The archaeology of beekeeping*. London (Duckworth).

Crane 1994: Crane, E. Beekeeping in the world of ancient Rome. *Bee World*, 75(3): 118-134.
Crane 1997: Crane, E., Look at this way. *Outlook on Agriculture*, 26(1): 3-5.
Crane 1999: Crane, E., *The world history of beekeeping and honey hunting*. London (Duckworth).
Crane 2003: Crane, E. *Making a bee-line. My journeys in sixty countries, 1949-2000*. Cardiff (IBRA).
Della Rocca 1790: Della Rocca, Ab. *Traité complet sur les abeilles*. III T., A Paris (De l'Imprimerie de Monsieur).
Davis 1996: Davis, J. L. A page turns in the history of Greek regional studies. *Journal of Roman Archaeology*, 9: 458-465.
Dermatopoulos 1954: Δερματόπουλος, Β. *Η πρακτική μελισσοκομία*. Αθήναι (Σπύρου).
Dermatopoulos 1984: Δερματόπουλος, Β. *Η σύγχρονη πρακτική μελισσοκομία*. Αθήνα (Σπύρου).
Dimitropoulos 1983: Δημητρόπουλος, Π. Η κατάσταση της μελισσοκομίας στο Ν. Μεσσηνίας. *Μελισσοκομική Ελλάς*, 33(418): 110-112.
Eckert 1943: Eckert, G. *Die Wanderbienenzucht in der Chalkidike*. Thessaloniki (N. Nikolaidis).
Eleftheriadis 1992: Ελευθεριάδης, Ε. *Λαογραφικά Λαραχανής της Ματσούκας του Πόντου*. Αθήνα (Καλλιτεχνικός Οργανισμός Ποντίων Αθηνών).
Efthymiou-Chatzilakou 1979/80: Ευθυμίου-Χατζηλάκου, Μ. Η καλαθοπλεκτική του Άργους. *Εθνογραφικά*, 2: 57-82.
Efthymiou-Chatzilakou 1981/82: Ευθυμίου-Χατζηλάκου, Μ. Μικρό σημείωμα για το μελισσοκόφινο. *Εθνογραφικά*, 3: 12-13.
Efstratiou 1984: Ευστρατίου, Ν. Εθνοαρχαιολογικές έρευνες στη Θράκη. *Αρχαιολογία*, 13: 20-26.
Faure 1973: Faure, P. *La vie quotidienne en Crète au temps de Minos* (1500 avant Jésus - Christ). Paris (Librairie Hachette).
Fielder 1840: Fielder, K. G. *Reise durch alle Theile des Königreiches Griechenland*. Leipzig (Freischer).
Fiorini 1880: Fiorini G. Journey to Cyprus and the East (translated from L' Apicoltore by Ch. Dadant). *American Bee Journal*, 16 (8): 373-377.
Francis 2016: Francis, J. E. Apiculture in Roman Crete. In: J. E. Francis, A. Kouremenos (eds), *Roman Crete: New Perspectives*. Oxford and Philadelphia (Oxbow Books), pp. 83-100.
Frantzeskakis 1903: Φραντζεσκάκης, Ε. Δ. *Πρακτική μελισσοκομία*. Εν Χανίοις (Εκ του τυπογρ. «Η Πρόοδος»).
Fraser 1950: Fraser, H. M. The story of the progress of beekeeping before 1800. *Bee World*, 31(5): 33-38.
Fraser 1951: Fraser, H. M. *Beekeeping in Antiquity*. London (University of London Press, 2nd ed., 1st ed. 1931).

Fraser 1955: Fraser, H. M. Beekeeping in the British Isles before 1500. *Bee World*, 36(5): 177-186.

Fraser 1958: Fraser, H. M. *History of beekeeping in Britain*. London (Bee Research Association).

Galt 1812: Galt, J. *Voyages and travels, in the years 1809, 1810 and 1811*. London (Cadell & Davies).

Gardikas 2015: Gardikas, K. The fragmented environment of interwar Halkidiki. In: Gounaris, B. C. (ed.), *Mines, olives and monasteries: Aspects of Halkidiki's environmental history*. Thessaloniki (Επίκεντρο & Pharos Books), pp. 163-183.

Georgandas 1957: Georgandas, P. D. The forerunner of the modern hive. *Bee World*, 38(11): 286-289.

Giohalas 2010: Γιοχάλας, Τ. Π. *Άνδρος: Αρβανίτες και Αρβανίτικα*. Αθήνα (Τυποθήτω).

Golding 1848: Golding, R. *The shilling bee book*. London (Longman, Brown, and Co, 2nd ed., 1st ed. 1847).

Gouridis 1967: Γουρίδης, Ι. Βρυσικά Διδυμοτείχου – Έβρου (Λαογραφική έρευνα). *Θρακικά*, 41: 224-280.

Guy 1971: Guy, R. A. commercial beekeeper's approach to the use of primitive hives. *Bee World*, 52(1): 18-24.

Harizanis 1995: Χαριζάνης, Π. Ικαρία, η μελισσοκομία και οι ιδιομορφίες της. *Μελισσοκομική Επιθεώρηση*, 9(10): 357-360.

Harissis 2018a: Χαρίσης, Χ. Β. Προϊστορικοί μελισσοκόμοι – αναζητώντας τα ίχνη τους στον ελλαδικό χώρο. *Αρχαιολογία & Τέχνες*, 126: 78-91.

Harissis 2018b: Harissis, H. V. Beekeeping in prehistoric Greece. In: Hatjina, F., Mavrofridis, G., Jones, R. (eds.), *Beekeeping in the Mediterranean from Antiquity to the Present*. Nea Moudania (HAO "Demeter"; Chamber of Cyclades; Eva Crane Trust), pp. 18-39.

Harissis 2020: Χαρίσης Χ. Β. Η μελισσοκομία στην προϊστορική και αρχαία Ελλάδα. In: Χατζήνα, Φ., Μαυροφρύδης, Γ. (eds), *Μελισσοκομία στη Μεσόγειο από την Αρχαιότητα ως Σήμερα*. Αθήνα (ΕΛΓΟ), pp. 18-39.

Harissis & Harissis 2009: Harissis, H. V., Harissis, A. V. *Apiculture in the prehistoric Aegean*. Oxford (Archaeopress, BAR International Series 1958).

Harissis and Mavrofridis 2012a: Harissis, H. V., Mavrofridis, G. A 17th century testimony on the use of ceramic top-bar hives. *Bee World*, 53(1): 56-58.

Harissis & Mavrofridis 2012b: Χαρίσης, Χ. Β., Μαυροφρύδης, Γ. Κυφέλες κινητής κηρήθρας στη βενετοκρατούμενη Κρήτη. Η μαρτυρία του Zuanne Papadopoli. *Μελισσοκομική Επιθεώρηση*, 26(4): 270-273.

Hatzopoulos 1977: Χατζόπουλος, Γ. Κ. *Συμβολή εις την λαογραφίαν του Κρυόνερου Ανατολικής Θράκης*. Τ. Α΄, Θεσσαλονίκη (Κοινότητα Καλαμπακίου Δράμας).

Hertz 1994: Hertz, O. The top-bar basket hive. *Bees for Development*, 33: 8-9.

Houliaras 2003: Χουλιαράς, Α. Μελισσοκομία. *Ευρυτανικά Νέα*, φύλλο 53 (19 Νοεμβρίου).

Hussein 2001: Hussein, M. H. Beekeeping in Africa. *Apiacta*, 36(1-2): 32-48

Johansson & Johansson 1972: Johansson, T. S. K., Johansson, M. P. Bee-library of the late Rev. L. L. Langstroth. *Bee World*, 53(1): 22-27.

Juvanec 2002: Juvanec, B. *Markovci, slamnati čebelnjak Slovenija*. Ljubljana (Univerza v Ljubljani).

Kambouroglou 1889: Καμπούρογλου, Δ. *Μνημεία της ιστορίας των Αθηναίων*. Τ. Α΄. Εν Αθήναις (Παπαγεωργίου).

Karastergios & Kokkora 2010: Καραστέργιος, Χ., Κόκκορα, Γ. Η περιοχή της Ιερισσού στα μέσα του 19^{ου} αιώνα με τα μάτια ενός περιηγητή. *Κύτταρο Ιερισσού*, 4: 10-11.

Katsaros 2011: Κατσαρός Θ. Βασίλης Τρικαλιώτης, ο άρχοντας των κοφινιών. *Μελισσοκομικό Βήμα*, 9(52): 29-31.

Katsouleas 2000: Κατσουλέας, Σ. Γ. Ο σχετικός με τη μέλισσα γλωσσικός και παροιμιακός πλούτος. Ο όρος κυψέλη. In: *ΣΤ΄ Τριήμερο Εργασίας, Η μέλισσα και τα προϊόντα της*, Νικήτη 12-15 Σεπτ. 1996. Αθήνα (ΠΤΙ ΕΤΒΑ), pp. 339-370.

Kazhdan & Epstein 2009: Kazhdan, P. A., Epstein, A-W. *Αλλαγές στον βυζαντινό πολιτισμό κατά τον11ο – 12ο αιώνα*. Αθήνα (ΜΙΕΤ, μτφ. Α. Παππάς).

Khatib 2003: Khatib, H. *Palestine and Egypt under the Ottomans. Paintings, Books, Photographs, Maps, Manuscripts*. New York (Tauris Parke).

Kigatiira 1976: Kigatiira, I. K. Keeping bees in fixed-comb and movable-comb frameless hives. In: *1st Conference "Apiculture in Tropical Climates"*, London 18-20 Oct. 1976. London (IBRA), pp. 9-13.

Kouvounas 2000: Κουβούνας, Θ. Τα πέτρινα μελισσοκομεία Ανατολής (Σελίτσανης) και Τσαριτσάνης. *Μελισσοκομική Επιθεώρηση*, 14(9): 397-400.

Kouvounas 2004: Κουβούνας, Θ. Παραδοσιακή μελισσοκομία. In: *2ο Επιστημονικό Συνέδριο Μελισσοκομίας – Σηροτροφίας*, Αθήνα 23-21 Μαΐου 2004, Ανακοινώσεις. Θεσσαλονίκη (ΕΕΕΜ-Σ), pp. 218-232.

Kouvounas 2005: Κουβούνας, Θ. Παραδοσιακή μελισσοκομία στο νομό Λάρισας. *Μελισσοκομική Επιθεώρηση*, 19(3): 167-171.

Kouvounas 2007: Κουβούνας, Θ. Οι δρόμοι των μελισσάδων. Παραδοσιακή μελισσοκομία στο τόξο του Βορείου Αιγαίου. In: *3ο Επιστημονικό Συνέδριο Μελισσοκομίας – Σηροτροφίας*, Θεσσαλονίκη 22-21 Απρ. 2007, Ανακοινώσεις. Θεσσαλονίκη (ΕΕΕΜ-Σ), pp. 270-276.

Kouvounas 2008: Κουβούνας, Θ. Οι δρόμοι των μελισσάδων. Στιγμιότυπα. *Μελισσοκομική Επιθεώρηση*, 22(3): 162-165.

Kouvounas 2021: Κουβούνας, Θ. Για το ξεχασμένο μελισσοκόφινο της Σκιάθου. *Μελισσοκομική Επιθεώρηση*, 35(277): 183-185 & 35(278): 251-264.

Krauss 2018: Krauss, S. *Studien zur byzantinisch-jüdischen Geschichte*. London (Forgotten Books, reprint of the 1914 Vienna edition).

Kritsky 2010: Kritsky, G. *The quest for the perfect hive. A history of innovation in bee culture.* New York (Oxford Univ. Press).

Kyrou 2000: Κύρου, Δ. Η μελισσοκομία στην οικονομία και τον καθημερινό βίο της Αρναίας σε παλαιότερες εποχές. In: *ΣΤ' Τριήμερο Εργασίας, Η μέλισσα και τα προϊόντα της*, Νικήτη 12-15 Σεπτ. 1996. Αθήνα (ΠΤΙ ΕΤΒΑ), pp. 371-389.

Kyrou 2005: Κύρου, Δ. Πληροφορίες για τις σχέσεις της Θάσου με τη Χαλκιδική και το Άγιον Όρος κατά τον 19° αιώνα και τις αρχές του 20ού αιώνα. *Θασιακά*, 12: 401-422.

Leontidis 1986: Λεοντίδης, Τ. *Τα κρητικά καλάθια. Μορφολογική, κατασκευαστική μελέτη.* Αθήνα (Μουσείο Κρητικής Εθνολογίας).

Liakos 2006: Λιάκος, Β. Η μελισσοκομία στη Σκύρο. *Μελισσοκομική Επιθεώρηση*, 20(5): 260-264.

Liapis 2007: Λιάπης, Α. Κ. *Καλάθια και καλαθάδες στη Χερσόνησο του Αίμου. Οι Ρομά και οι άλλοι.* Αλεξανδρούπολη (Οικουμενικότης).

Manga 1995: Manga, S. Baskets can be used for TBH. *Bees for Development*, 37: 5.

Marcenay 1979: Marcenay, Ph. *L'home et l'abeille.* Paris (Berger-Levrault).

Mavrogenis 1979a: Μαυρογένης, Γ. Μελισσοκομική τεχνική της μινωικής και της αρχαίας ελληνικής εποχής στο νησί της Κρήτης. *Μελισσοκομική Ελλάς*, 29(367): 20-21.

Mavrogenis 1979b: Μαυρογένης, Γ. Εγχώριες κυψέλες της δυτικής Κρήτης. *Μελισσοκομική Ελλάς*, 29(374-375): 244-245.

Mavrofridis 2006: Μαυροφρύδης, Γ. Η μελισσοκομία στον μινωικό – μυκηναϊκό κόσμο. *Μελισσοκομική Επιθεώρηση*, 20(5): 268-272.

Mavrofridis 2007a: Μαυροφρύδης, Γ. Προλήψεις και παραδόσεις των Ελλήνων μελισσοκόμων. *Μελισσοκομική Επιθεώρηση*, 21(1): 24-28.

Mavrofridis 2007b: Μαυροφρύδης, Γ. Τα κοφίνια του Έβρου. *Μελισσοκομική Επιθεώρηση*, 21(2): 79-80.

Mavrofridis 2007c: Μαυροφρύδης, Γ. Μελισσοκομία με κινητές κηρήθρες. Μια διαχρονική πρακτική με αρχαίες καταβολές. In: *3ο Επιστημονικό Συνέδριο Μελισσοκομίας – Σηροτροφίας*, Θεσσαλονίκη 22-21 Απρ. 2007, Ανακοινώσεις. Θεσσαλονίκη (ΕΕΕΜ-Σ), pp. 132-163.

Mavrofridis 2007d: Μαυροφρύδης, Γ. Νέα στοιχεία για την παραδοσιακή μελισσοκομεία των Κυθήρων και Αντικυθήρων. *Μελισσοκομική Επιθεώρηση*, 21(3): 161-163.

Mavrofridis 2008a: Μαυροφρύδης, Γ. Σε αναζήτηση της κινητής κηρήθρας. Νήσοι Αργοσαρωνικού. *Μελισσοκομική Επιθεώρηση*, 22(3): 166-170.

Mavrofridis 2008b: Μαυροφρύδης, Γ., Τα ελληνικά επίστομα κοφίνια. *Μελισσοκομική Επιθεώρηση*, 22(5): 320-324.

Mavrofridis 2009a. Μαυροφρύδης, Γ. Ελληνικές κυψέλες με ισομεγέθεις κηρηθροφορείς πριν την ανακάλυψη της ΚΤΒΗ. *Μελισσοκομική Επιθεώρηση*, 23(4): 288-291.

Mavrofridis 2009b: Μαυροφρύδης, Γ. Η μελισσοκομία των Πομάκων της Ροδόπης. *Μελισσοκομική Επιθεώρηση*, 23(5): 346-350.

Mavrofridis 2009c: Μαυροφρύδης, Γ., Οι μελισσοκομικές πανηγύρεις της Χαλκιδικής. *Μελισσοκομική Επιθεώρηση*, 23(6): 396-399.

Mavrofridis 2010a: Μαυροφρύδης, Γ. Τα ταξίδια των Δυτικών για αγορές μελισσιών στην Κύπρο τον 19° αιώνα. *Μελισσοκομική Επιθεώρηση*, 24(1): 34-39.

Mavrofridis 2010b: Μαυροφρύδης, Γ. Σε αναζήτηση της κινητής κηρήθρας II. Οι «ελληνικές κυψέλες» στα έργα περιηγητών και ερευνητών. *Μελισσοκομική Επιθεώρηση*, 24(2): 106-111.

Mavrofridis 2010c: Μαυροφρύδης, Γ. Τα οριζόντια κοφίνια. *Μελισσοκομική Επιθεώρηση*, 24(4): 276-281.

Mavrofridis 2010d: Μαυροφρύδης, Γ. Σε αναζήτηση της κινητής κηρήθρας III. Αργολίδα. *Μελισσοκομική Επιθεώρηση*, 24(5): 348-352.

Mavrofridis 2010e: Μαυροφρύδης, Γ. Σε αναζήτηση της κινητής κηρήθρας IV. Τροιζηνία. *Μελισσοκομική Επιθεώρηση*, 24(6): 428-431.

Mavrofridis 2011a: Μαυροφρύδης, Γ. Η μελισσοκομία στην αρχαία Αίγυπτο I. *Μελισσοκομική Επιθεώρηση*, 25(1): 52-56.

Mavrofridis 2011b: Μαυροφρύδης, Γ. Αναφορικά με την προέλευση των ελληνικών επίστομων κοφινιών. *Μελισσοκομική Επιθεώρηση*, 25(3): 208-213.

Mavrofridis 2011c: Μαυροφρύδης, Γ. Η μελισσοκομία στον ρωμαϊκό κόσμο. *Μελισσοκομική Επιθεώρηση*, 25(4): 266-271.

Mavrofridis 2012a: Μαυροφρύδης, Γ. Τα κοφίνια κινητής κηρήθρας. *Μελισσοκομική Επιθεώρηση*, 26(3): 174-178.

Mavrofridis 2012b: Μαυροφρύδης, Γ. Η μελισσοκομία στην Αττική στα τέλη του 18ου αιώνα. *Μελισσοκομική Επιθεώρηση*, 26(6): 400-404.

Mavrofridis 2013a: Μαυροφρύδης, Γ. Τα «κλεμπούρια» της Κέρκυρας. *Μελισσοκομική Επιθεώρηση*, 27(1): 33-35.

Mavrofridis 2013b: Μαυροφρύδης, Γ. Κυψέλες κινητής κηρήθρας στην αρχαία Ελλάδα. *Αρχαιολογική Εφημερίς*, Τ. 152: 15-27.

Mavrofridis 2013c: Μαυροφρύδης, Γ. Τρύγος με θανάτωση των μελισσών στην παραδοσιακή μελισσοκομία. *Μελισσοκομική Επιθεώρηση*, 27(3): 158-160.

Mavrofridis 2013d: Mavrofridis, G. Experimental Archaeology. Beekeeping with copies of ancient upright hives. *Bee World*, 90(4): 82-84.

Mavrofridis 2014a: Μαυροφρύδης, Γ. Οι παραδοσιακές κυψέλες του Πόντου. *Μελισσοκομική Επιθεώρηση*, 28(3): 185-188.

Mavrofridis 2014b: Μαυροφρύδης, Γ. Οι παραδοσιακές κυψέλες της Μικράς Ασίας. *Μελισσοκομική Επιθεώρηση*, 28(4): 260-264.

Mavrofridis 2014c: Μαυροφρύδης, Γ. Μελισσοκομικές γνώσεις και πρακτικές των χρηστών κυψελών κινητής κηρήθρας του 17ου και 18ου αιώνα. *Μελισσοκομική Επιθεώρηση*, 28(6): 411-414.

Mavrofridis 2015a: Μαυροφρύδης, Γ., Η νομαδική μελισσοκομία πριν την έλευση της σύγχρονης κυψέλης. *Μελισσοκομική Επιθεώρηση*, 29(241): 176-179.

Mavrofridis 2015b: Μαυροφρύδης, Γ., Η παραδοσιακή μελισσοκομία της Αρναίας (19ος – 20ός αι.). *Αρναία*, 28(109): 12-18.

Mavrofridis 2016a: Μαυροφρύδης, Γ. Η αναπαραγωγική μελισσοκομία της Χαλκιδικής. *Μελισσοκομική Επιθεώρηση*, 30(245): 36-39.

Mavrofridis 2016b: Μαυροφρύδης, Γ. Μελισσομαντριά. *Μελισσοκομική Επιθεώρηση*, 30(247): 196-200.

Mavrofridis 2016c: Μαυροφρύδης, Γ. Η παραδοσιακή μελισσοκομία στην Ιερισσό. *Κύτταρο Ιερισσού*, 13: 10-11.

Mavrofridis 2017a: Μαυροφρύδης, Γ. Οι παραδοσιακές κυψέλες κινητής κηρήθρας. *Πελοποννησιακά Γράμματα*, Τ. 2: 299-334.

Mavrofridis 2017b: Μαυροφρύδης, Γ. Το νομαδικό κοφίνι της βόρειας Ελλάδας. *Μελισσοκομική Επιθεώρηση*, 31(251): 41-45.

Mavrofridis 2017c: Μαυροφρύδης, Γ. Η παραδοσιακή μελισσοκομία της Θάσου. *Θασιακά*, 18: 267-299.

Mavrofridis 2017d: Μαυροφρύδης, Γ. Κηρόμυλοι και μελόπρεσες. *Μελισσοκομική Επιθεώρηση*, 31(252): 125-129.

Mavrofridis 2017e: Μαυροφρύδης, Γ. Η επιρροή των ελληνικών κυψελών στην εξέλιξη της παγκόσμιας μελισσοκομίας. *Μελισσοκομική Επιθεώρηση*, 31(253): 189-193.

Mavrofridis 2017f: Μαυροφρύδης, Γ. Θασιακή μελισσοκομία (16ος – 20ός αι.). *Μελισσοκομική Επιθεώρηση*, 31(256): 432-437.

Mavrofridis 2017g: Μαυροφρύδης, Γ. Σοφίας Γερμανίδου Βυζαντινός μελίρρυτος πολιτισμός (βιβλιοκρισία). *Μελισσοκομική Επιθεώρηση*, 31(256): 459-460.

Mavrofridis 2017h: Mavrofridis, G. Traditional wax and honey presses of southeastern Europe. *Ethnoentomology*, 1: 74-83.

Mavrofridis 2018a: Μαυροφρύδης, Γ. Ελληνιστικά πώματα κυψελών για προστασία των μελισσών από τη *Vespa orientalis*. In: *Proceedings of the 9th International Scientific Meeting on Hellenistic Pottery*, Thessaloniki Dec. 5-9 2012. Τ. II. Athens (ΤΑΠΑ), pp. 849-856.

Mavrofridis 2018b: Μαυροφρύδης, Γ. Μελισσοκομία στον ελληνορωμαϊκό κόσμο – οι κυψέλες. *Αρχαιολογία & Τέχνες*, 127: 100-111.

Mavrofridis 2018c: Μαυροφρύδης, Γ. Παραδοσιακή μελισσοκομία. *Αρχαιολογία & Τέχνες*, 128: 66-79.

Mavrofridis 2018d: Μαυροφρύδης, Γ. Τα αχυρένια μελισσοκόφινα. *Μελισσοκομική Επιθεώρηση*, 32(262): 416-421.

Mavrofridis 2019a: Μαυροφρύδης, Γ. Μελισσοκομικά περιβάλλοντα στο Αθηναϊκό Πεδίο: από την παραδοσιακή μελισσοκομία στη σύγχρονη παραγωγή. *Μελισσοκομική Επιθεώρηση*, 33(264): 122-126.

Mavrofridis 2019b: Μαυροφρύδης, Γ. *Η ελληνική παραδοσιακή μελισσοκομία και η συμβολή της στις διεθνείς εξελίξεις*. Αθήνα (ΙΓΕ).

Mavrofridis 2019c: Μαυροφρύδης, Γ. Παραδοσιακή μελισσοκομία στις Θρακικές Σποράδες. *Μελισσοκομική Επιθεώρηση*, 33(265): 156-160.

Mavrofridis 2019d: Mavrofridis, G. Traditional beekeeping in Crete (17th – 20th century). In: *Proceedings of the 12th International Congress of Cretan Studies*, Heraklion, 21-25 Sept. 2016. Heraklion (SCHS), pp. 1-15.

Mavrofridis 2020: Μαυροφρύδης, Γ. Παραδοσιακή μελισσοκομία με κυψέλες κινητής κηρήθρας στη δυτική Κρήτη. *Εν Χανίοις*, Τ. 14: 397-420.

Mavrofridis 2021a: Mavrofridis, G. Feeding bees (*Apis mellifera* and *Apis cerana*) with poultry meat through the ages. *Bee World*, 28(2): 68-70.

Mavrofridis 2021b: Μαυροφρύδης, Γ. Οι ελληνικοί κηρόμυλοι. *Ο Μυλολόγος*, 6: 6-8.

Mavrofridis 2022: Mavrofridis, G. A new approach to the study of ancient Greek Beekeeping. In: Wallace-Hare, D. (ed.), *New Approaches to the Archaeology of Beekeeping*. Oxford (Archaeopress), pp. 1-18.

Mavrofridis & Anagnostopoulos 2012: Mavrofridis, G., Anagnostopoulos, I. Th. The first top-bar hive with fully interchangeable combs. *American Bee Journal*, 152(5): 483-485.

Mavrofridis & Chairetakis 2019: Μαυροφρύδης Γ., Χαιρετάκης, Γ. Η μελισσοκομία της Σαλαμίνας στη διαχρονία. *Μελισσοκομική Επιθεώρηση*, 33(263): 40-44.

Mavrofridis & Goutzamani 2019: Μαυροφρύδης, Γ., Γκουτζαμάνη, Μ. Μέλισσα και μελισσοκομία στα Γεωπονικά. *Μελισσοκομική Επιθεώρηση*, 33(268): 411-415.

Mavrofridis & Petanidou 2022a: Μαυροφρύδης, Γ., Πετανίδου, Θ. Παραδοσιακή μελισσοκομία με κινητές κυψέλες στο νησί της Άνδρου. *Μελισσοκομική Επιθεώρηση*, 36(284): 271-276.

Mavrofridis & Petanidou 2022b. Μαυροφρύδης, Γ., Πετανίδου, Θ. Λιθόκτιστοι μελισσότοιχοι με θυρίδες για την προστασία παραδοσιακών κυψελών. *Μελισσοκομική Επιθεώρηση*, 36(286): 397-402.

Mavrofridis & Tselios 2013: Μαυροφρύδης, Γ., Τσέλιος, Χ. Η παραδοσιακή μελισσοκομία του Νομού Σερρών. *Μελισσοκομική Επιθεώρηση*, 27(2): 87-90.

Mavrofridis et al. 2021: Μαυροφρύδης, Γ., Τάταρης, Γ., Πετανίδου, Θ. Η γεωγραφία της παραδοσιακής νομαδικής μελισσοκομίας στα νησιά του Αιγαίου. *Γεωγραφίες*, 37: 21-34.

Mavrofridis et. al. 2022a: Μαυροφρύδης, Γ., Τάταρης, Γ., Πετανίδου, Θ. Η γεωγραφία της παραδοσιακής νομαδικής μελισσοκομίας της Χαλκιδικής *Αρναία*. 35(136): 20-28.

Mavrofridis et. al. 2022b: Μαυροφρύδης, Γ., Τάταρης, Γ., Τσιλιγγίρη, Ε., Πετανίδου, Θ. Παραδοσιακή μελισσοκομία και λιθόκτιστες κατασκευές στο νησί της Άνδρου. *Επετηρίς Εταιρείας Κυκλαδικών Μελετών*, 24: 631-649.

Mavrofridis et. al. forthcoming: Mavrofridis, G., Tataris, G., Petanidou. T. Traditional migratory beekeeping in Greece, 18th – mid 20th century. *Journal of Apicultural Research* (Accepted on 09 April 2023).

Melanofrydis 1955: Μελανοφρύδης, Π. Η. Ακ-Νταγκ. *Ποντιακή Εστία*, 6(61): 3035-3037.

Melas 1999: Melas, M. The ethnography of Minoan and Mycenaean beekeeping. In: Betancourt, P. P., Karageorgis, R., Laffineur, R., Niemeir, W-D (eds), *Meletemata. Studies in Aegean Archaeology Presented to Malcolm H. Wiener as he Enters his 65th Year. Aegaeum 20*, Vol. II. Liège – Austin (Université de Liège), pp. 485-491.

Melena 2014: Melena, J. L. Mycenaean writing. In: Duhoux, Y., Morpurgo Davis, A. (eds), *A companion to Linear B: Mycenaean Greek texts and their world*. Vol. 3. Louvain la Neuve – Walpole (Peeters), pp. 1-186.

Mestre 2003: Mestre, J-R. La vannerie apicole. *Les Cahiers d'Apistoria*, 1: 64-72.

Mitsi 2003: Μήτση, Ε. Στη σκιά των μνημείων. Αρχαιολάτρες και αρχαιόσυλοι στην Ελλάδα του 18ου αιώνα. *Σύγχρονα Θέματα*, 60(82): 60-67.

Nicolaidis 1955: Nicolaidis, N. J. Facts about beekeeping in Greece. *Bee World*, 36(8): 141-149.

Nixon & Moody 2017: Nixon, L., Moody, J. Cultural landscapes and resources in Sphakia, SW Crete: A diachronic perspective. In: *From Maple to Olive, Proceedings of a Colloquium to Celebrate the 40th Anniversary of the Canadian Institute in Greece*, Athens, 10-11 June 2016. Athens (The Canadian Institute in Greece), pp. 485-504.

Owens 2018: Owens, G. Η φωνή του Δίσκου της Φαιστού. Διάλεξη στο Εθνικό Κέντρο Τεκμηρίωσης (ΕΚΤ) στις 7 Φεβρουαρίου 2018. Το βίντεο της διάλεξης υπάρχει στον ιστότοπο του ΕΚΤ (http://helios-eie.ekt.gr/EIE/handle/10442/15651).

Pallis 2009: Πάλλης, Γ. *Τοπογραφία του αθηναϊκού πεδίου κατά τη μεταβυζαντινή περίοδο*. Θεσσαλονίκη (Κέντρο Βυζαντινών Ερευνών).

Pallis 2019: Πάλλης, Γ. Η Μονή Δαφνίου στις δαγγεροτυπίες του Joseph Philipert Girault de Pringey (1842). In: *Ανασκαφή και Έρευνα XII, Από το ερευνητικό έργο του Τομέα Αρχαιολογίας και Ιστορίας της Τέχνης, Δωδέκατο Επιστημονικό Συνέδριο*, Αθήνα 28-29 Μαρτ. 2019. Περιλήψεις. Αθήνα (ΕΚΠΑ), pp. 88-89.

Papadopoulo 1965: Papadopoulo, P. The Greek basket hive is cheap and efficient. *Shell Farmer*, 1: 4-6.

Papadopoulo 1967: Papadopoulo, P. Beekeeping. *Bee-keeping*, 4: 1-8.

Papadopoulo 1969: Papadopoulo, P. Rhodesia, her bees and beekeepers. *Apiacta*, 4(2): 16-22.

Papadopoulos 2010: Παπαδόπουλος, Χ. Παραδοσιακή μελισσοκομία. Ανακοίνωση στην Ημερίδα για τη Μέλισσα που διοργάνωσε η Ένωση Πολιτιστικών Φορέων Έβρου στις 2010/02/27 και καταγράφηκε σε δίσκο ακτίνας.

Papageorgiou 2016: Papageorgiou, I. Truth lies in the details: Identifying an apiary in the miniature wall painting from Akrotiri, Thera. *The Annual of the British School at Athens*, 111: 95-120.

Papaggelos 2000: Παπάγγελος, I. Η μελισσοκομία στη Χαλκιδική κατά τους μέσους χρόνους και την τουρκοκρατία. In: *ΣΤ' Τριήμερο Εργασίας, Η μέλισσα και τα προϊόντα της*, Νικήτη 12-15 Σεπτ. 1996. Αθήνα (ΠΤΙ ΕΤΒΑ), pp. 190-210.

Papaioannou 1939: Παπαϊωάννου Χ. Γεωργοοικονομική έρευνα της κωμοπόλεως Αρναίας. *Δελτίον Αγροτικής Τραπέζης της Ελλάδας*, 2: 146-187.

Papaoikonomou 2012: Παπαοικονόμου, Ν. Ε. Η κτηνοτροφία της Λιαρίγκοβης το 1845. *Αρναία*, 25(94): 16-17.

Papaoikonomou 2014: Παπαοικονόμου, Ν. Ε. Ένα οθωμανικό φορολογικό τεφτέρι του 1845 από τη Λιαρίγκοβη. *Αρναία*, 27(105): 11-12.

Papavasileiou 2008: Παπαβασιλείου, Ε. Η μόνιμη κατοίκηση στα νεότερα χρόνια. In: *Το Φαράγγι της Σαμαριάς*. Χανιά (Φορέας Διαχείρισης Εθνικού Δρυμού Σαμαριάς), pp. 93-128.

Paterson 2008: Paterson, P. Memories of Eva Crane. In: *Eva Crane. Bee Scientist 1912-2007*. Cardiff (IBRA), pp. 97-99.

Pattinson 2012: Pattinson, D. Pre-modern beekeeping in China: A short history. *Agricultural History*, 16(4): 235-255.

Petropoulos 1957: Πετρόπουλος, Δ. Α. Μελισσοκομικά Χαλκιδικής και Δυτικής Μακεδονίας. *Λαογραφία*, 17(Α'): 186-196, 356-357.

Polemis 1981: Πολέμης, Δ. *Ιστορία της Άνδρου*. Άνδρος (Καΐρειος Βιβλιοθήκη).

Price & Nixon 2005: Price, S., Nixon, L. Ancient Greek Agricultural Terraces: Evidence from texts and archaeological survey. *American Journal of Archaeology*, 109(4): 665-694.

Primovski 1960: Примовски, А. Село Бабяк, Разлоско. In: *Езиковедско – етнографски изледвания в памет на Академик Стоян Романски*. София (БАН), pp. 619-650.

Primovski 1973: Примовски, А. *Бит и култура на Родопските Българи*. Сборник за Народни Умотворения и Наропис LIV. София (БАН).

Raichevski 1996: Райчевски, С. Пчеларство. In: *Страиджа, Материална и духовна култура*. София ("Проф Марин Дринов"), pp. 81-83.

Rammou & Bikos 2000: Ράμμου Αικ., Μπίκος Θ. Η Ελλάδα της μελισσοκομίας. Τρία χρόνια μελισσοκομικών καταγραφών. In: *ΣΤ' Τριήμερο Εργασίας, Η μέλισσα και τα προϊόντα της*, Νικήτη 12-15 Σεπτ. 1996. Αθήνα (ΠΤΙ ΕΤΒΑ), pp. 423-435.

Richards-Mantzoulinou 1979: Richards-Μαντζουλίνου, Ε. Μέλισσα Πότνια. *Αρχαιολογικά Ανάλεκτα εξ Αθηνών*, 12: 72-89.

Rizopoulou-Igoumenidou 2000: Ριζοπούλου-Ηγουμενίδου, Ε. Η παραδοσιακή μελισσοκομία στην Κύπρο και τα προϊόντα της (μέλι, κερί) κατά τους νεότερους χρόνους. In: *ΣΤ' Τριήμερο Εργασίας, Η μέλισσα και τα προϊόντα της*, Νικήτη 12-15 Σεπτ. 1996. Αθήνα (ΠΤΙ ΕΤΒΑ), pp. 390-408.

Rotroff 2006: Rotroff, S. I. *The Athenian Agora, Volume 33. Hellenistic pottery: The pain ware*. Princeton (American School of Classical Studies at Athens).

Ruttner 1979a: Ruttner, F. Minoische und altgriechische Inkertechnik auf Kreta. In: *Bienenmuseum und Geschihte der Bienenzucht, Internationales Symposium über Bienenwirtschaft*, Freiburg 16-18 Aug. 1977. Bukarest (Apimondia), pp. 209-229.

Ruttner 1979b: Ruttner, F. *Evolution historique de la ruche*. Bucarest (Apimondia).

Savvakis 1994: Σαββάκης, Κ. Ιστορική εξέλιξη της κυψέλης στη μελισσοκομία της Κρήτης. *Μελισσοκομική Επιθεώρηση*, 8(5): 175-180.

Sellianakis 1998: Σελλιανάκης, Β. Ιστορία και εξέλιξη της μελισσοκομίας στη νήσο Κρήτη. Unpublished thesis at the Agricultural University of Athens.

Shinas 1887: Σχινάς, Ν. *Οδοιπορικαί σημειώσεις. Μακεδονίας τεύχος Γ΄*. Εν Αθήναις (Messager d' Athenes).

Simopoulos 1990: Σιμόπουλος, Κ. *Ξένοι ταξιδιώτες στην Ελλάδα*. 4 Τ., Αθήνα (χ.ε., 7th ed., 1st ed. 1973).

Smyrnakis 1903: Σμυρνάκης, Γ. *Το Άγιον Όρος*. Εν Αθήναις (Ελληνισμός).

Speis 2016: Speis, G., *Beekeeping on the Island of Andros: An ethnographic approach*. Andros (Kaireios Library; Eva Crane Trust).

Spencer 1973: Spencer T. *Fair Greece, sad relic: Literary philhellenism from Shakespeare to Byron*. New York (Octagon, 2nd ed., 1st ed. 1954).

Spon 1678: Spon, J. *Voyage d'Italie, de Dalmatie, de Grèce, et du Levant, fait aux années 1675 & 1676*. III T. A Lyon 1678 (Antoine Cellier les fils).

Stathaki-Koumari 1985: Σταθάκη-Κούμαρη, Ρ. *Η καλοθοπλεκτική στην Ελλάδα*. Αθήνα (ΕΟΜΜΕΧ).

Stojianov 1942: Стояновъ, Н. Природата на островъ Тасосъ. *Беломорски Прегледъ*, I: 73-106.

Stouraitis 1907: Στουραΐτης Ι. Η μελισσοκομία εν Λεβαδεία. *Μελισσοκομική Εφημερίς*, 2(4): 2-3.

Thompson 1744: Thompson, Ch. *The travels of the Late Charles Thompson Esq*. III V. Reading (J. Newberry & C. Micklewrighe).

Topalidis 1940: Τοπαλίδης, Ν. Οι μελισσοτροφικές συνθήκες της νήσου Θάσου, *Θρακική Μελισσοκομία*, 2(19-20): 159-162.

Toufexis 1904: Τουφεξής, Γ. Λ. *Οδηγός μελισσοκόμου*. Εν Αθήναις (Ι. Δ. Κολλάρος).

Toufexis 1909: Τουφεξής, Γ. Λ. *Μελισσοκομία*. Εν Αθήναις (Εκ του τυπογρ. Νομικής).

Tozer 1890: Tozer, H. F. *The islands of the Aegean*. Oxford (Clarendon Press).

Tsellios 1988: Τσέλλιος Δ. Νομαδική μελισσοκομία, χθες και σήμερα. *Μελισσοκομική Επιθεώρηση*, 2(6): 175-176.

Tsellios 1996: Τσέλλιος, Δ. Το κοφίνι της Χαλκιδικής. *Μελισσοκομική Επιθεώρηση*, 10(5): 184-187.

Tsilogeorgis 2010: Τσιλογεώργης, Ν. Το μετόχι της Μονής Ιβήρων στο Κάστρο. *Θασιακά*, 10: 695-715.

Tsiriktzidou 1989: Τσιρικτζίδου, Σ. Τα γαλλικό εμπόριο κατά τον 18° αιώνα στη Θάσο. *Θασιακά*, 6: 123-129.

Typaldos-Xydias 1927: Τυπάλδος-Ξυδιάς, Α. *Η νομαδική μελισσοκομία εν Ελλάδι*. Αθήναι (Παράρτημα Γεωργικού Δελτίου).

Typaldos-Xydias 1981: Τυπάλδος-Ξυδιάς, Α. Νομαδική μελισσοκομία. *Μελισσοκομική Ελλάς*, 31(391): 8-12.

Tyree et al. 2012: Tyree, L., Robinson, H. L., Stamataki, P. Minoan bee smokers: An experimental approach. In: Mantzourani E., Betancourt, P. (eds), *Philistor, Studies in Honor of Costis Davaras*. Philadelphia (INSTAP Academic Press), pp. 223-232.

Urquhart 1838: Urquhart, D., *The spirit of the East*. II Vol., London (Colburn).

Vakarelski 1977: Вакарелски, Х. *Етнография на България*. София (Наука и Изкуство).

Vendenabeele & Olivier 1979: Vendenabeele, F., Olivier, J-P. *Les idéogrammes archéologiques du Linéaire B*. Paris (Ètudes crétoises XXIV, Ècole Française d'Athènes).

Vigopoulou 2005: Βιγοπούλου, Ι. *Η ανάδυση και η ανάδειξη του ελληνισμού στα ταξίδια των περιηγητών (15°ς – 20°ς αιώνας)*. Ανθολόγιο από τη Συλλογή Δημητρίου Κοντομηνά. Κατάλογος Έκθεσης (16 Μαΐου - 16 Ιουνίου 2005). Αθήνα (Κότινος).

Vrontis 1938/1948: Βρόντης, Α. Η μελισσοκομία και το μαντατόρεμα στη Ρόδο. *Λαογραφία*, 12: 195-230.

Walpole 1817: Walpole, R. *Memoirs relating to European and Asiatic Turkey*. London (Longman, Hurst, Rees, Orme and Brown, 2nd ed. 1818).

Wheler 1682: Wheler G. *A journey into Greece*. London (W. Cademan).

Whitfield 1956: Whitfield, B. G. Virgil and the bees. A study in ancient apicultural lore. *Greece and Rome*, Series 2, 3(2): 99-117.

Wildman 1770: Wildman, T. *A treatise on the management of bees*. London (W. Strahan & T. Cadell, 2nd ed., 1st ed. 1768).

Yfantidis 1997: Υφαντίδης, Μ. *Μελισσοκομία, Επιστήμη και εφαρμογή*. Θεσσαλονίκη (χ.ε., 3rd ed., 1st ed. 1983).

Zachos, G. 2021: Ζάχος, Γ. Κείων μελιτουργίαν. In: *Έξοχος άλλων, Τιμητικός τόμος για την καθηγήτρια Εύα Σημαντώνη-Μπουρνιά*. Αθήνα (Οργανισμός Διαχείρισης και Ανάπτυξης Πολιτιστικών Πόρων).

Zymbragoudakis 1979: Zymbragoudakis, Ch. The bee and beekeeping of Crete. *Apiacta*, 14(3): 134-138.

Georgios Mavrofridis studied archaeology at Sofia University (Master of Arts) and economics at the University of West Attica. He is a PhD candidate at the Department of Geography, University of the Aegean, having earned a scholarship from the State Scholarship Foundation.

For almost two decades now, he has been dealing with issues related to archaeology, ethnography, and the history of beekeeping. He is very involved in ongoing on-site ethnographic research, collecting information from old beekeepers and recording traditional equipment and stone beekeeping structures. At the same time, he practices experimental beekeeping by using copies of ancient Greek and Byzantine ceramic hives to clarify in practice the methods used in the past.

He has published more than 150 articles in journals and collective volumes, three book chapters and has co-edited two books. He has also published the following books (in Greek): *The Greek Traditional Beekeeping and its Contribution to International Developments* (Athens 2019), *Greek Woven Hives* (Athens 2021), *Greek Wicker Hives from Antiquity to the Present* (Thessaloniki 2022). Recently, he has co-authored *The Beekeeping of Kea Island. From Antiquity to the Present* (Athens 2023).

www.ingramcontent.com/pod-product-compliance
Lightning Source LLC
Chambersburg PA
CBHW041243240426
43670CB00024B/2968